Terahertz Time Domain Spectroscopy

Tunan Chen · Dexian Yan · Kai Zhong ·
Longhai Liu · Jining Li · Jianquan Yao

Terahertz Time Domain Spectroscopy

Technology and Applications

SSTP

Springer

Tunan Chen
Department of Neurosurgy
Southwest Hospital
Chongqing, China

Kai Zhong
Tianjin University
Tianjin, China

Jining Li
Tianjin University
Tianjin, China

Dexian Yan
China Jiliang University
Hangzhou, Zhejiang, China

Longhai Liu
Tianjin University
Tianjin, China

Jianquan Yao
Tianjin University
Tianjin, China

ISBN 978-981-95-2604-8 ISBN 978-981-95-2605-5 (eBook)
https://doi.org/10.1007/978-981-95-2605-5

Jointly published with Shanghai Scientific and Technical Publishers
The print edition is not for sale in China (Mainland). Customers from China (Mainland) please order the print book from: Shanghai Scientific and Technical Publishers.
ISBN of the co-publisher's edition: 978-7-5478-5954-4

The original submitted manuscript has been translated into English. The translation was done using artificial intelligence. A subsequent revision was performed by the author(s) to further refine the work and to ensure that the translation is appropriate concerning content and scientific correctness. It may, however, read stylistically different from a conventional translation.

This Springer imprint is published by the registered company Springer Nature Singapore Pte Ltd.
The registered company address is: 152 Beach Road, #21-01/04 Gateway East, Singapore 189721, Singapore

Acknowledgement Authors would like to acknowledge the contributions from below researchers.

Fei Li
Hua Feng
Ning Mu
Yu Chen
Meng Zhou
Kang Ma
Yang Yang
Chuanyan Yang
Chenfu Guo
Yaning Shi
Zhenliang Xu
Lixia Duan
Ying Lai
Kai Xu
Pengfei Zhou
Weina Li
Haijun Duan
Yulian Quan
Jingmin Zheng
Pengnan Zhao
Bo Chen
Qian Chen
Hongfeng Guo
Ping Zhao

Competing Interests The authors have no competing interests to declare that are relevant to the content of this manuscript.

Contents

About the Authors

Tunan Chen received his Ph.D. from the Third Military Medical University in 2012. He is currently Associate Chief Physician and Associate Professor at the First Affiliated Hospital of Army Medical University. His primary research focuses on optical diagnosis and treatment of gliomas. In 2021, he received the honorary title as a Chongqing Talent (Young Top Talent). He received the inaugural Outstanding Young Neuroscientist Award from the Continuing Education Committee of the National Health Commission in 2024.

Dexian Yan received his Ph.D. from Tianjin University, China and is currently an Associate Professor at China Jiliang University. His research focuses on laser and optoelectronics, as well as microwave and terahertz technologies. He has conducted extensive work in areas such as nonlinear optical frequency conversion, microwave and terahertz waveguides, micro/nano metamaterial devices, and plasma-based applications.

Prof. Dr. Kai Zhong is from School of Precision Instrument and Opto-electronics Engineering, Tianjin University, China. His research focuses on laser and terahertz technology. He was granted the title of a Young Talent of the Innovative Talents Promotion Plan of Tianjin City, Jin Guofan Prize for Excellent Youth of China Instrument Society Scholarships, Tianjin 131 Innovative Talents Program, etc.

Longhai Liu received the Doctor of Engineering degree in Advanced Manufacture from Tianjin University, China. He works with Advantest (China) Co., Ltd., Shanghai, China. His research interests include terahertz time-domain spectroscopy and its nondestructive testing. He has been awarded the First Prize of Henan Provincial Science and Technology Achievement Awards in 2024.

Assoc. Prof. Jining Li received Ph.D. in Optics from Nankai University, China. From 2010 to 2012, he had an overseas visiting experience in T-ray Group with Prof. W. Withayachumnankul and Prof. A. Derek at School of Electrical and Electronic Engineering, University of Adelaide, Australia. In 2015, he joined the Faculty of

School of Precision Instrument and Opto-electronics Engineering, Tianjin University, China. His research interests include laser technology and application, THz technology and functional devices based on metamaterials, for example all-solid-state laser, laser detection technology, the physical interaction mechanism of THz wave and novel materials, high-performance active THz functional devices and the applications in biomedical, and THz target imaging detection technology. He is a Committee Member of Basic Medical Imaging Branch of Tianjin Medical Imaging Technology Research Association, and a Council Member of the Laser Society of Tianjin and the Optical Society of Tianjin.

Jianquan Yao received the master's degree in Precision Instrument, Tianjin University, China, in 1965. Since 1966, he has been with Tianjin University, as an Assistant Professor and a Lecturer, where he is a Professor and the Director of the Institute of Laser and Opto-Electronics. He was a Visiting Scholar with the Department of Applied Physics, Stanford University, Stanford, CA, USA, from 1980 to 1982. In 1997, he was elected as an Academician with Chinese Academy of Sciences, Beijing, China. He has long been engaged in research of all-solid laser, nonlinear optics frequency conversion, and terahertz science and technology. Professor Yao has been awarded the World Gold Medal of Invention, Eureka, Brussels, and the National Invention Award Class II of China.

Chapter 1
Introduction

1.1 Concept of Terahertz Waves

Terahertz waves are defined aselectromagnetic waves in the frequency range of 0.1 THz to 10 THz, corresponding to a wavelength range of 3–30 μm. Terahertz waves occupy the region microwave and infrared regions of the electromagnetic spectrum, overlapping with the extremely high frequency microwave band at lower frequencies and the far-infrared band at higher frequencies. At 1 THz (1×10^{12} Hz), the electromagnetic wave oscillation period is 1 ps, the wavelength is 300 μm, and the wavenumber is 33 cm^{-1}. The photon energy at 1 THz is 4.1 meV, corresponding to a thermodynamic temperature of 48 K. Numerous natural objects emit thermal radiation within the terahertz range, making these waves ubiquitous in the ambient environment.

To date, research on terahertz waves has been ongoing for several decades. Initially, due to the lack of effective methods to generate and detect these waves, researchers found it challenging to conduct in-depth studies of this band, dubbed the "terahertz gap" in electromagnetic wave research. When the oscillation frequency of the electromagnetic field reaches the terahertz range, the stray capacitance effect—negligible at lower frequencies—becomes increasingly significant, limiting classical electromagnetic wave radiation sources and electronic devices based on macroscopic charge motion from generating terahertz waves. For high-frequency terahertz waves, optical radiation typically originates from electronic quantum transitions, where the energy difference between two levels corresponds to terahertz photons of merely a few millielectronvolts (meV). This energy is smaller than that of most crystal optical phonons and is therefore significantly affected by thermal relaxation effects. Researchers primarily investigate terahertz waves from both electronic and optical approaches.

T. Chen et al., *Terahertz Time Domain Spectroscopy*,
https://doi.org/10.1007/978-981-95-2605-5_1

1.2 Research and Characteristics of Terahertz Waves

Since terahertz waves are situated between the microwave and infrared bands, they combine characteristics of both microwaves and infrared light. Moreover, terahertz waves exhibit unique properties absent in microwaves and infrared light.

The pulse width of terahertz waves is typically in the order of picoseconds, providing picosecond and sub-picosecond time resolution, enabling the study of transient spectra. Systematic sampling measurement techniques can effectively suppress background noise interference from the surrounding environment, and time-domain spectroscopy measures the real-time power of terahertz pulses through coherent measurement rather than average power, resulting in a high signal-to-noise ratio.

Terahertz waves can effectively penetrate many dielectric materials and non-polar liquids. In practical applications, they complement microwaves and X-rays, making them suitable for non-destructive testing in security inspections or quality monitoring. Additionally, the wavelength of terahertz waves is much larger than the size of particles such as dust and smoke in the air, resulting in significantly less scattering compared to visible and infrared light. This allows terahertz waves to transmit effectively even under adverse conditions like fires or sandstorms. Polar molecules generally absorb terahertz waves strongly, such as water. However, this strong absorption characteristic can be utilized to measure and distinguish different states of the same substance or the content of different components. Notably, due to its properties, water can also serve as a potential terahertz source. Firstly, compared to generating terahertz waves in air (with a nonlinear polarization rate of 10^{-19} cm^2/W, neutral atom density of 10^{16} cm^{-3}, oxygen ionization energy of approximately 12.1 eV, and nitrogen ionization energy of approximately 15.6 eV), water exhibits superior physical properties (with a nonlinear polarization rate of 10^{-16} cm^2/W in the visible range and 10^{-10} cm^2/W in the terahertz range, neutral atom density of 10^{21} cm^{-3}, and ionization energy of approximately 6.5 eV). Secondly, unlike traditional crystal materials, water does not require phase matching conditions and does not face phonon absorption issues caused by crystal oscillations. Its renewable and circulating flow characteristics make it highly suitable for constructing stable terahertz sources. Furthermore, the photon energy of terahertz waves is relatively low, with a photon energy of only 4.1 meV for a terahertz wave at 1 THz, which is lower than the energy of various chemical bonds. This ensures that measurements using terahertz waves do not ionize the sample, making the process safer.

The rotational and vibrational energy level transitions of many molecules, especially organic molecules, correspond to the terahertz range. The vibrational energy levels of dielectrics, semiconductor materials, superconducting materials, and thin film materials also fall within the terahertz frequency range. This makes terahertz waves an effective tool for studying material properties, enabling identification and characterization of materials. Compared to microwaves, terahertz waves possess shorter wavelengths and higher frequencies, enabling higher bandwidth in communication systems and improved signal transmission directionality.

1.3 Applications of Terahertz Technology

With the rapid development of terahertz radiation sources and detection technologies, the excellent properties of terahertz waves are increasingly recognized by researchers, showing great potential in fields such as physics, materials science, biology, and astronomy. Currently, the applications of terahertz technology mainly include terahertz spectroscopy, terahertz security inspection, terahertz communication, and terahertz radar, as illustrated in Fig. 1.1.

1.3.1 Terahertz Spectroscopy Applications

In past decades, the rapid development of ultrafast laser technology has provided reliable and stable laser sources for generating terahertz pulses. The increasing maturity of research on terahertz sources, detectors, and transmission lines has facilitated the rapid development of terahertz wave technology. Terahertz Time Domain Spectroscopy (THz-TDS) is the most fundamental and important technology in terahertz applications, complementing traditional infrared and Raman spectroscopy

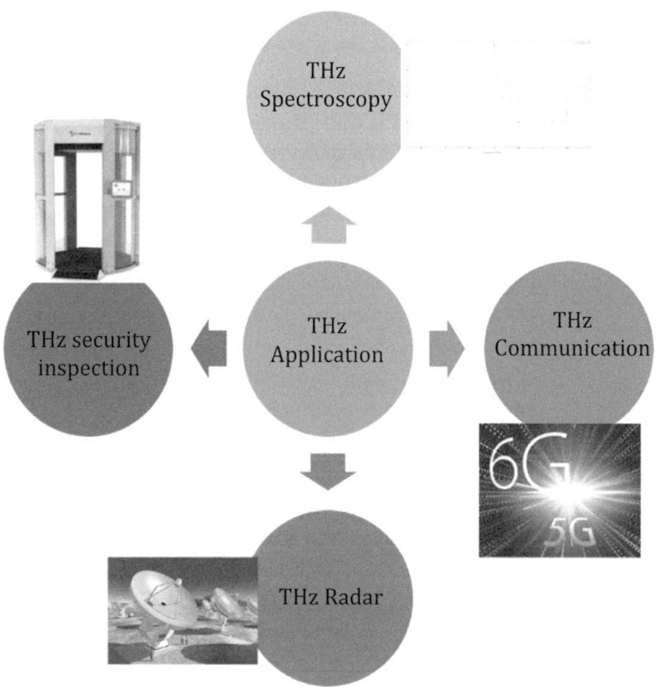

Fig. 1.1 Applications of terahertz technology

techniques. THz-TDS is an emerging coherent detection technology and a fundamental and widely applied area of terahertz technology research. It has become an essential tool for studying material properties in the terahertz frequency range. By measuring the terahertz time-domain pulse signals of both the air background and the sample, THz-TDS simultaneously obtains the amplitude and phase of the terahertz pulse signals. Through fast Fourier transform and related calculations of the terahertz time-domain waveform, it can derive spectral parameters such as transmittance, reflectance, phase, power, absorption coefficient, refractive index and permittivity of the sample. THz-TDS shows great potential for development in studying the spectral characteristics of materials in the terahertz range, investigating the vibrational and rotational properties of constituent molecules, analyzing material composition and structure, and measuring material thickness. THz-TDS will become a powerful tool for revealing and analyzing ultrafast phenomena in fundamental sciences such as physics, chemistry, and biology. Spectral imaging technology based on THz-TDS will also have broad application fields and significant development potential. It has important applications in crystal optics, food and agricultural products, biomedicine, non-destructive testing, and cultural heritage preservation. These topics will be introduced in detail in subsequent chapters.

1.3.2 Terahertz Security Screening Applications

Enhancing security checks in crowded places such as airports, stations, and exhibitions is an important means of preventing dangerous incidents. Security screening technology is one of the key technologies to be developed in the field of public safety. Current human security screening technologies mainly include X-ray transmission imaging, X-ray backscatter imaging, non-imaging radar, terahertz imaging based on passive imaging principles, and millimeter-wave holographic imaging technology. X-ray transmission imaging uses X-rays to create transmission images of the human body, requiring X-rays to penetrate the body, which involves relatively high doses and poses significant safety risks, often used in special situations such as detecting internally concealed drugs. X-ray backscatter imaging forms images by receiving X-rays scattered back from the body surface, significantly reducing the dose compared to transmission imaging and having strong detection capabilities for contraband hidden on the body surface. However, due to the ionizing nature of X-rays, this technology remains controversial, and its promotion is still restricted.

The application of terahertz technology in security screening has enormous commercial value, attracting the attention of researchers in related fields both domestically and internationally due to its immense potential and market prospects. Europe and America started research earlier, forming a relatively complete and mature terahertz technology system, and have pioneered the development of advanced terahertz security imaging systems. Active terahertz imaging systems use high-power coherent light and ultra-low noise heterodyne detection technology. Millimeter-wave body imaging technology is currently an advanced technology in the global security field, used for passenger security screening at airports in countries such as the United States, the United Kingdom, the Netherlands, Australia, and Japan. The United States and the European Union have established access technology standards. Terahertz security imaging equipment can effectively detect items concealed under clothing on various parts of the body without direct contact, especially non-metallic items, and can obtain information on the shape, size, and location of concealed items from images. Additionally, millimeter-wave body imaging equipment is harmless to the human body, has strong penetration capabilities, and its emission power is less than one-thousandth of that of mobile phone electromagnetic radiation. It can accurately identify items carried by individuals, effectively improving the objectivity, accuracy, and specificity of inspections, reducing the labor intensity of security personnel, and enhancing screening efficiency.

Currently, the terahertz industry system in Europe and America start the industrialization process accelerating, and many high-tech companies have been established. Companies such as VDI in the United States, Teratech in the United Kingdom, and FARRAN in Ireland mainly produce related terahertz components; companies like ThruVision in the UK, Millivision, and Brijot in the US can provide terahertz imaging security imaging systems. In recent years, mergers and acquisitions among terahertz technology companies in Europe and America have been common, with strong companies entering this emerging market, such as the American security and defense industry giant L3 Communications. Its ProVision imaging systems is shown in Fig. 1.2. It operates at a frequency of around 35 GHz. Using active scanning holographic imaging technology, this equipment can complete a scan in 2 s, with a resolution at the sub-centimeter level [1].

The terahertz sparse array imaging radar from Rohde & Schwarz uses a square sparse array scheme, operating at a frequency around 80 GHz. It enables real-time imaging of large target field areas, with an imaging time of approximately 32 ms and

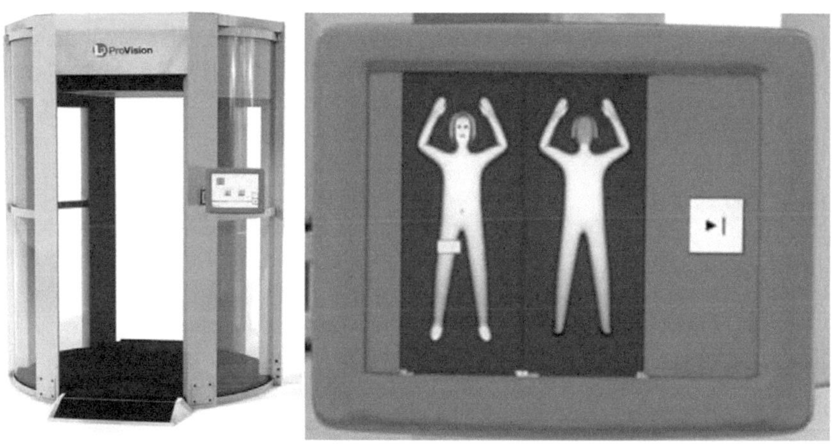

Fig. 1.2 ProVision security system by L3 communications

a resolution of about 3 mm. The system employs a multi-transmit and multi-receive method to achieve coherent reception of terahertz radiation backscattered from the human body. By using a sparse array approach, the number of transceiver elements required by the system is reduced from 25,600 to around 700. The MMW-Imager-90 imaging system from France's MC2 company uses a one-dimensional linear detector array combined with one-dimensional mechanical scanning to achieve rapid imaging of target areas. Its imaging speed can reach over 10 frames per second, allowing for quick imaging of moving targets. The Millivision uses a passive detector array combined with quasi-optical scanning to achieve rapid imaging of the human body, also at a speed of 10 frames per second. The ThruVision series of terahertz imaging systems from the UK operates around 250 GHz, capable of passive imaging of targets at about 15 m, but with poor resolution and image quality, making it unable to image small volume targets.

In recent years, several countries including the United States, Germany, and Finland have begun researching terahertz imaging devices based on low-temperature superconducting bolometers. The VTT Research Center in Finland developed a 0.1–1.2 THz ultra-wideband superconducting microbolometer terahertz imaging system, the 350 GHz superconducting TES junction terahertz imaging system developed by the U.S. National Institute of Standards and Technology, and the 350 GHz superconducting terahertz imaging system reported by the German Institute of Photonic Technology. These types of terahertz security inspection devices have an imaging range of twenty to thirty meters, making it difficult to achieve a large field of view. Most use superconducting detector arrays to achieve passive imaging of targets by collecting thermal radiation. The use of superconducting detection requires operation in a low-temperature environment of 4 K or below, making mass production challenging.

With breakthroughs in key core technologies, the development of domestically produced terahertz chips, modules, and other components is accelerating, particularly in the application of terahertz body security inspection. In 2014, the first domestic passive terahertz body security inspection system was developed. This system requires 2 s for imaging, with a resolution of 2 cm. The core key technologies of passive terahertz imaging security inspection systems, such as terahertz detectors, optical systems, and post-processing algorithms, have broken through foreign technology blockades, which is of great significance. Subsequently, the country invested substantial research funding to conduct research in related fields,

To date, domestically, high-definition, short-focus, wide-angle, and long-focus passive imaging terahertz security inspection systems have been developed, applicable in different scenarios. High-definition security inspection systems can achieve an imaging resolution of 1 cm, capable of detecting small contraband items carried by individuals; short-focus security inspection devices are compact, with a short imaging distance and small footprint, suitable for confined spaces; wide-angle security inspection devices can achieve an imaging field of 2 m long and 1.6 m wide, allowing simultaneous security checks for multiple people; long-focus security inspection devices can image targets 4 m away, with a compact body, convenient for mobile or wall-mounted installation.

Subsequently, various passive terahertz imaging body security inspection systems have been developed domestically for different applications, including channel-type, rotating, and vehicle-mounted mobile systems. Channel-type security inspection systems can perform non-radiative, zero-contact, continuous rapid security checks, easily applicable in high-traffic areas such as subways, airports, and stations; rotating security inspection systems require the person being inspected to rotate in place for precise detection from different angles; vehicle-mounted mobile security inspection systems offer high mobility, allowing flexible installation for one-stop inspection, applicable at conference sites [2].

In recent years, to better implement new technology applications, the Civil Aviation Administration's Public Security Bureau has actively explored the trial evaluation of new technologies at security checkpoints. In August 2017, the bureau organized and provided policy support for the simultaneous on-site trial evaluation of millimeter-wave body imaging equipment at four airports: Beijing Capital, Shanghai Hongqiao, Shanghai Pudong, and Qingdao Liuting. In May 2018, the bureau conducted a throughput pressure test of millimeter-wave body imaging equipment at Nanjing Lukou and Qingdao Liuting airports, further verifying the equipment's throughput efficiency while ensuring safety. The next step will involve synchronously conducting research and pilot projects on new models and policies for "differentiated security checks," aiming to continuously innovate security check models, optimize processes, and improve efficiency while enhancing security, thereby creating a better travel experience for passengers. In June 2018, the Civil Aviation Administration issued the "Internal Control Standards for the Appraisal of Civil Aviation Millimeter-Wave Body Imaging Security Inspection Equipment" and the "Testing Procedures for the Detection Capability of Prohibited Items by Civil Aviation Millimeter-Wave Body Imaging Security Inspection Equipment," officially incorporating millimeter-wave body imaging equipment into the list of security inspection equipment for Chinese civil aviation. This made China the third country globally and the first in Asia to independently issue standards for millimeter-wave body imaging technology. In the future, millimeter-wave body imaging equipment will gradually replace metal detection gates at civilian airports, providing passengers with a safer and more efficient personal security check service. On May 28, 2021, the GA/T 1779-2021 "General Technical Requirements for Passive terahertz Imaging Body Security Inspection Equipment" standard was released and implemented on October 1, 2021. This standard was proposed and managed by the National Security and Alarm System Standardization Technical Committee (SAC/TC100) The standard document specifies the general technical requirements for passive terahertz imaging body security inspection equipment, describing the packaging, marking, storage and transportation, accompanying documents, test methods, and establishing inspection rules for such equipment. This standard document applies to the design, manufacture, inspection, and use of passive terahertz imaging body security inspection equipment. As the COVID-19 pandemic spreaded globally, the need for normalized epidemic prevention and control led to the integration of temperature measurement and information recognition functions in passive Terahertz security systems. This allows for the acquisition of body temperature information while ensuring security detection, and combined with big data analysis, it can determine activity trajectories and enhance security checks and epidemic prevention. In the same year, to prepare for and ensure the safety of the Asian Games, the Zhejiang Provincial Key Research and Development Program—Non-Radiative Human terahertz Imaging Rapid Security Inspection Instrument Development Project was approved, contributing to the development of terahertz security inspection instruments.

1.3.3 Terahertz Communication Applications

With the large-scale commercial application of fifth-generation mobile communication (5G), global research and exploration of sixth-generation mobile communication technology (6G) have begun. Looking towards 2030 and beyond, human society will enter an era of intelligence. From mobile interconnection to the Internet of Everything, and then to the intelligent interconnection of everything, 6G will achieve a transition from serving people and objects to supporting the efficient connection of intelligent entities, through human–machine-object intelligent interconnection and collaborative symbiosis, meeting the high-quality development requirements of the economic society and contributing to the construction of a digital society. Terahertz communication and visible light communication are important components of 6G communication, with terahertz communication mainly applied in short-distance, high-capacity communication scenarios. On June 6, 2021, the IMT-2030 (6G) Promotion Group officially released the "6G Overall Vision and Potential Key Technologies" white paper, as a phased achievement of the promotion group. The content covers the overall vision, eight major business application scenarios, ten potential key technologies, and some thoughts on 6G development. Regarding the 6G spectrum resource issue, it is proposed that 6G will expand to higher frequency bands and efficiently utilize low, medium, and high full-spectrum resources. Among them, low-frequency spectrum will remain a strategic resource for 6G development, millimeter waves will play a more important role in the 6G era, and higher frequency bands such as terahertz will focus on meeting the short-distance, high-capacity needs of specific scenarios. The terahertz band, with its rich spectrum resources and unique properties, has attracted significant attention from major world powers and received strong support from the International Telecommunication Union, making it a highly potential key candidate spectrum technology for 6G.

Terahertz communication, compared to traditional microwave point-to-point communication, increases communication frequency to achieve faster transmission rates and larger operational bandwidths. Based on domestic and international validation and research results of terahertz communication systems, the operational bandwidth of terahertz communication is mostly above 2 GHz, with achievable maximum transmission rates of 100 Gbit/s. However, the propagation and penetration losses of terahertz waves also increase correspondingly. With constant transmission and reception gain in the communication system, the communication distance using terahertz waves will significantly decrease. Compared to the microwave band, terahertz waves have relatively poor penetration and diffraction capabilities, making terahertz communication unsuitable for long-distance communication but more suitable for short-distance point-to-point transmission. Due to the current scarcity of spectrum resources and the surge in wireless communication devices, terahertz communication also needs to enhance communication capacity from different dimensions.

Terahertz communication can serve as an effective supplement to existing air interface transmission methods, with significant application potential in scenarios such as holographic communication, inter-chip communication, nano-communication, ultra-large capacity data backhaul, and short-range ultra-high-speed transmission. Additionally, high-precision positioning and high-resolution sensing based on terahertz communication signals are also important application areas.

(1) Terahertz Ground Wireless Communication

Terahertz communication can be used in ground wireless communication systems. Terahertz wireless transceivers can replace fiber optics or cables for high-speed backhaul of base station data, enabling high-speed data transmission in areas where fiber optics cannot be used. Since terahertz communication supports bandwidth and rates superior to millimeter-wave communication, it has significant application potential in the field of fixed wireless access. With the rapid development of information and communication technology, the demand for cloud service applications has increased significantly, intensifying the need for data center applications. Traditional data center architecture uses cables for connections, which occupy space and incur high maintenance costs. Terahertz, with its ultra-high communication rates, has potential application value in wireless data centers, reducing space costs, cable maintenance costs, and power consumption. The high path loss, short propagation distance, and good directivity of terahertz communication make it suitable for secure access between devices, ultra-high-speed secure downloads, and secure payments. This application has been defined in the IEEE Std 802.15.3d-2017 standard, with a high level of technical standardization maturity [3]. With advancements in wireless mobile communication technology, terahertz communication can provide ultra-high-speed network coverage for hotspot areas, complementing macro cell networks and offering ultra-broadband wireless communication for small areas. Future miniaturized, low-power, and low-cost terahertz devices can be used to implement terahertz wireless local area networks, enabling 6G communication services. Terahertz communication can achieve high-speed links between nearby devices and can also be used in terahertz wireless personal area network systems for wireless links between personal electronic devices.

In May 2020, the research group led by Xianbin Yu at Zhejiang University integrated fiber optic communication with terahertz communication, proposing a multi-carrier modulation signal processing mechanism based on probabilistic shaping to enhance system bit energy efficiency. By combining linear-nonlinear equalization and phase noise compensation algorithms to overcome system interference constraints, they were the first internationally to achieve long-distance photonic terahertz communication transmission at rates above 100Gbit/s in the 350 GHz band, improving the efficiency of photonic terahertz communication systems and expanding the wireless access range of amplifier-free terahertz communication [4].

In August 2021, South Korea's LG Electronics conducted joint research with Fraunhofer-Gesellschaft, a leading European applied research institution, in Berlin, Germany. They successfully transmitted terahertz wireless communication in an outdoor environment over a range of 100 m, demonstrating terahertz data transmission using adaptive beamforming and high-gain antenna switching technology, achieving stable communication in the 155–175 GHz frequency range. LG Electronics anticipates that 6G communication will be commercialized by 2029 and states that standardization will begin in 2025.

At the 2022 Geneva International Invention Exhibition, the team led by Professor Xiaohu You from the Purple Mountain Laboratories stood out with their project "100/200 GbE Photonic terahertz Real-time Seamless Integrated Communication System for 6G," winning the gold medal. This system first proposed a real-time terahertz-fiber integrated transmission architecture, achieving a record-breaking single-channel 103.125 Gb/s and dual-channel 206.25 Gb/s high-speed real-time 2×2 MIMO wireless communication in the typical terahertz frequency range of 360–430 GHz for 6G communication. By adopting a novel photonic-assisted terahertz technology route and leveraging commercially mature digital coherent optical modules, it breaks through the bottleneck of electronic devices and addresses the real-time processing challenges of traditional terahertz wireless communication signals. This technology integrates the high bandwidth and large capacity of fiber optic communication with the portable and flexible access of terahertz wireless communication, significantly lowering the R&D threshold for 6G and greatly accelerating the commercialization process of terahertz technology.

The Guangdong Province New Generation Communication and Network Innovation Research Institute, in collaboration with the School of Information Science and Technology at Fudan University and the Aerospace Information Research Institute of the Chinese Academy of Sciences, has conducted research on 6G terahertz wireless communication technology. The research team has developed a 200 GHz all-solid-state electronic terahertz system, which can be used for a complete physical layer transmission system for broadband terahertz wireless communication. By employing an innovative adaptive bit power loading discrete multitone modulation technique, the system's usable bandwidth and spectral efficiency are enhanced, achieving an average spectral efficiency of 6.1bit/s/Hz across the full bandwidth. Ultimately, a single-channel transmission rate exceeding 130Gbps was achieved with a signal bandwidth of 21.32 GHz.

In March 2022, the research group led by Yu Xianbin at Zhejiang University, in collaboration with DTU, UC3M, KTH, RISE, Shanghai Jiao Tong University, and other domestic and international units, achieved injection locking of a photonic integrated dual-wavelength light source based on an optical frequency comb, generating optoelectronic terahertz signals with a hertz-level linewidth and a frequency tunable from 0 to 1.4 THz. By combining high spectral efficiency modulation formats and advanced digital signal processing technology, they successfully broke through the 400 GHz band with a single-channel rate exceeding 100 Gbit/s for high-energy-efficiency terahertz wireless communication, with power consumption per bit distance as low as 2.9E-18 J/bit/m, providing a technical reserve for on-chip integrated optoelectronic terahertz communication [5].

(2) Terahertz Space Communication

Terahertz communication can also be applied in space communication systems. Terahertz waves can achieve lossless transmission in outer space, enabling long-distance transmission and avoiding interference from Earth's radiation noise. When high-speed aircraft enter the atmosphere, shock waves can generate high temperatures, ionizing the air and creating a plasma sheath around the aircraft, with a frequency around 60–70 GHz. Traditional measurement and communication methods find it difficult to penetrate the plasma sheath [6]. Terahertz wave frequencies are much higher than the plasma sheath frequency, allowing them to penetrate the plasma sheath for communication and measurement of the aircraft. Additionally, terahertz wavelengths are shorter than microwave wavelengths, and if terahertz antenna systems can be miniaturized and planarized, terahertz communication systems can be deployed on platforms such as satellites, drones, and airships, serving as wireless communication and relay equipment. They can be applied in satellite clusters, between space and ground, and for interstellar high-speed communication over distances exceeding thousands of kilometers, achieving integrated communication across air, space, ground, and sea.

(3) Micro-scale Terahertz Communication

Terahertz communication can be used in micro-scale communication systems. With shorter wavelengths, Terahertz waves hold the potential for realizing transmission and reception devices and components of millimeter size or even smaller, enabling ultra-high-speed data transmission over very short distances. With the rise and development of functional materials technology such as graphene, Terahertz communication can be applied not only on a traditional macro scale but also as a wireless nano-network communication band for high-speed data transmission on and between chips [7]. This includes wearable or implantable terahertz devices for nano-sensor networks and other micro-scale communication applications, bridging the gap from macro to micro communication in 6G networks. Micro-scale communication exhibits significant 6G characteristics, requiring the integration of Terahertz technology with micro-nano technology to break through new materials and process technologies, achieving millimeter-scale, efficient, and low-cost terahertz communication systems.

1.3.4 Terahertz Radar Applications

In recent years, with the rapid advancement of technologies for generating, transmitting, and detecting terahertz waves, the terahertz band has emerged as a new strategic high ground in military technology competition, and terahertz radar systems have been continuously developed [8]. Compared to microwave radar, terahertz radar has a shorter wavelength and larger bandwidth, providing high "space–time-frequency" resolution: it offers high imaging resolution in space, allowing observation of both coarse and fine structures of targets, enabling precise measurement of target features; it provides high imaging frame rates in time, facilitating real-time imaging of targets and guiding weapon systems for precise strikes; in the frequency domain, it offers Doppler sensitivity, aiding in micro-motion measurement and high-precision speed prediction. Additionally, terahertz radar has a narrow beam, enhancing antenna gain and angular tracking accuracy; its wide frequency range allows effective anti-jamming capabilities; and its compact size enables installation on small drones and their swarms, satellites, missiles, etc. Compared to laser radar, terahertz waves can penetrate smoke, dust, sand, and other environments, and are insensitive to the aerodynamic optical effects and thermal environment effects of high-speed moving targets, making them suitable for operations in complex environments and detection of high-speed moving targets. In 2012, the U.S. Defense Advanced Research Projects Agency (DARPA) conducted a video synthetic aperture radar program, with the radar operating in the frequency range of 231.5–235 GHz, using a one-transmit-four-receive antenna mode, and designed with an imaging frame rate of 5 Hz. In 2017, the program announced that it had successfully obtained real-time, full-motion video images of ground targets obscured by clouds using this system.

Since its inception, terahertz radar has primarily been used for early warning detection of space or ground military targets. As early as 2016, the United States initiated research on integrated terahertz radar technology for both terrestrial and space applications. By coordinating terahertz radars deployed on the ground and in space with traditional ground-based radars, it effectively counters fifth-generation stealth aircraft that rely on coatings and shape for stealth. In missile defense interception, terahertz active radar seekers, either independently or in combination with infrared, can serve as an effective means of warhead identification. Additionally, its high-precision ranging and speed measurement capabilities enable high-precision guidance of maneuvering warheads, unaffected by celestial or ground clutter. Terahertz radar can also be mounted on airships or satellites for detecting hypersonic targets in near space, penetrating plasma to image the target body from a distance, providing high-resolution body images.

In civilian applications, it can first be applied in the automotive radar field. During real driving conditions, especially on slippery roads in rainy weather or on highways with low visibility due to smog, chain collisions may occur. Using terahertz radar, potential collisions between the vehicle being driven and the vehicle ahead or obstacles can be detected, warning the driver to avoid danger. Compared to 24 GHz and 77 GHz millimeter-wave automotive radars, terahertz radar offers advantages in resolution and miniaturization. In 2015, the University of Birmingham in the UK developed a terahertz vehicle collision avoidance radar using frequency-modulated continuous wave signals, with a center frequency of 150 GHz and a bandwidth of 6 GHz, achieving a scanning imaging resolution of 3 cm $\times$ 38 cm.

Secondly, the terahertz radar can be used for meteorological cloud measurement and early warning of heavy rain and typhoons. Compared to millimeter-wave radar, terahertz radar can more accurately provide cloud particle characteristics, and space-borne platforms avoid the severe atmospheric attenuation of terahertz waves near the ground. Cloud remote sensing radar is evolving from the 94 GHz to above 200 GHz frequency bands. Since 1989, 215, 200, and 225 GHz radars have been used for monitoring studies of thin clouds, particles, and snowfall processes. Beijing Institute of Technology has conducted research on cloud particle scattering characteristics and designed and developed a terahertz cloud measurement radar system.

Additionally, based on the micro-Doppler sensitivity advantage, terahertz radar has a very broad prospect in non-contact life signal monitoring, avoiding the limitations of contact detection equipment. It can monitor the breathing and heartbeat of patients with extensive burns, infants, and moving crowds within surveillance areas, identifying safety hazards. In military applications, it can monitor the heartbeat and breathing of soldiers on distant battlefields, allowing command centers to promptly grasp battlefield casualty situations and provide references for subsequent combat plans. Currently, 120 and 228 GHz terahertz life signal monitoring radars have been developed and are undergoing indoor experiments.

1.4 Prospects for Terahertz Technology Development

1.4.1 Hardware Enhancement of THz-TDS Technology

THz-TDS systems mainly include ultrashort pulse laser pump sources, terahertz radiation antennas, terahertz wave detection antennas, and time delay control systems. The relatively low terahertz radiation power remains a major limitation for the application of THz-TDS technology in non-destructive testing. During actual spectral measurement and imaging, several issues may arise, potentially preventing experimental testing. These issues include:

(1) The system's optical path is complex, with numerous optical components, making it easily affected by external environmental influences, resulting in poor stability. Frequent system optimization is required, and daily maintenance is challenging, significantly limiting application and commercialization development.
(2) During spectral measurement or imaging, if the sample being measured is large or cannot be moved, the system cannot measure or image it, limiting practical applications.
(3) In certain counter-terrorism operations, experimental measurements need to be completed in relatively dangerous extreme environments. Experimenters often cannot move or approach the sample, potentially preventing sample measurement.
(4) The low conversion efficiency of optical terahertz sources, low sensitivity of detection devices, and system noise severely affect the signal-to-noise ratio and imaging contrast of terahertz signals. The detection part of the system typically

uses a lock-in amplifier for signal acquisition, increasing the system's size and cost.

Increasing the power of terahertz signals to enhance the penetration depth of terahertz waves and improve the signal-to-noise ratio of measurements.

1.4.2 Establishment of a Terahertz Spectral Database

In the field of terahertz spectroscopy detection, international institutions have conducted research on spectral databases for different materials. The National Institute of Standards and Technology (NIST) in the United States has published terahertz spectra for certain foods and drugs; the European "THz-Bridge" project has provided terahertz spectra for biomolecules such as proteins; in 2007, the RIKEN and the National Institute of Information and Communications Technology (NICT) in Japan established an open terahertz spectral database (http://thzdb.org), which is expanded annually, facilitating research and use of terahertz spectral analysis by various institutions; Shizuoka University in Japan has established a terahertz spectral database for pharmaceuticals (https://www.rie.shizuoka.ac.jp/~thz/shizuoka_univ_sasaki_lab_database/index.html), which is still being updated in real-time. Foreign research institutions have established various types of terahertz spectral databases in their respective research fields, but due to differences in samples, testing equipment, and testing environments, there is no unified standard. Common measurement systems include THz-TDS systems and far-infrared Fourier spectrometers, each with unique features in terms of bandwidth, resolution, and signal-to-noise ratio. In China, Tsinghua University, Capital Normal University, and Zhejiang University have reported terahertz spectral databases, but there is a lack of a unified and recognized database. Joint efforts from the industry are needed. Existing databases can effectively study information on spectral absorption peaks, but full-spectrum information is often required for identifying actual samples. Therefore, in specific application fields and spectral measurement systems, if the sample material to be studied is not in the existing spectral database, or if the existing data cannot meet actual needs, a new spectral database needs to be established.

1.4.3 Integration with Chemometrics Algorithms

In addition to researching terahertz spectra, different data analysis algorithms are used as auxiliary means to analyze the terahertz spectra of mixtures, improving the signal-to-noise ratio of the spectra, thereby enabling accurate qualitative and quantitative identification of mixed samples. These algorithms can mainly be divided into two categories: analytical algorithms for qualitative and quantitative analysis of data, and processing algorithms for data denoising and reconstruction.

In data analysis algorithms, researchers typically use deep learning or linear regression algorithms for qualitative and quantitative analysis of the samples under test. Mainstream algorithms include Partial Least Squares (PLS) regression, Support Vector Machine (SVM) algorithms, and Support Vector Regression (SVR) algorithms.

For actual mixed samples, interference from water vapor and other factors often results in a low signal-to-noise ratio of the spectra. Therefore, before using analytical algorithms, processing algorithms are needed to improve the signal-to-noise ratio of the spectra. Mainstream data processing algorithms include genetic algorithms, Principal Component Analysis (PCA), and denoising reconstruction algorithms.

1.4.4 Terahertz Near-Field for Enhanced Imaging Resolution

Due to the low photon energy of terahertz waves, they possess non-destructive and non-ionizing properties, allowing for widespread use in material detection and non-destructive testing fields. Terahertz imaging, particularly in the field of biomedical imaging, has attracted significant attention from researchers. Visible light cannot effectively penetrate samples, and X-ray contrast cannot meet requirements, whereas terahertz imaging can address both demands. Traditional far-field terahertz imaging systems are limited by the diffraction limit, with a minimum resolution of only the order of $\lambda/2$, meaning that the minimum resolution corresponding to 1THz is 0.15 mm. Millimeter-level imaging resolution somewhat restricts the application of terahertz imaging technology. Therefore, the development of near-field detection and microscopy techniques can help achieve higher imaging resolution. To overcome the diffraction limit and enhance spatial resolution, Germany's Protemics has proposed a terahertz near-field imaging system, enabling sub-wavelength resolution for terahertz point-by-point imaging. This work effectively enhances the performance of terahertz near-field imaging technology, providing a new pathway for terahertz imaging research.

Terahertz wavelengths are generally around 300 μm, and due to the diffraction limit, the optical spatial resolution of terahertz far-field measurement systems is generally limited to about 150 μm. The accuracy of THz light far-field measurement results often cannot meet the needs of material science research, especially for studies of nanoscale material distribution. The emergence of the THz-NeaSNOM near-field optical microscope provides an excellent solution to this problem. Germany's neaspec GmbH, in collaboration with Fraunhofer IPM, has successfully developed an easy-to-use experimental device with a terahertz system spatial resolution reaching 30 nm, based on the neaspec company's neaSNOM near-field optical microscope.

References

1. Yue A, Wu S. Development status of passive terahertz imaging human security inspection equipment. China Secur Prot Technol Appl. 2021;(5):15–18.
2. Lei T. Overview of the development and application of terahertz human security inspection technology. China Secur Prot Technol Appl. 2021;5:11–4.
3. IEEE Std 802.15.3d-2017, IEEE Standard for High Data Rate Wireless Multi-Media Networks—Amendment 2: 100 Gb/s Wireless Switched Point-to-Point Physical Layer, Approved 28 September 2017, IEEE-SA Standards Board.
4. Wang S, Lu Z, Li W, et al. 26.8-m THz wireless transmission of probabilistic shaping 16-QAM-OFDM signals. APL Photonics. 2020; 5(5):056105.
5. Jia S, Lo MC, Zhang L, et al. Integrated dual-laser photonic chip for high-purity carrier generation enabling ultrafast terahertz wireless communications. Nat Commun. 2022;13(1):1–8.
6. China Unicom Terahertz Communication Rechnology White Paper, June 2020
7. Chen Y, Han C. Channel modeling and characterization for wireless networks-on-chip communications in the millimeter wave and terahertz bands. IEEE Trans Mol Biol Multi Scale Commun. 2019;5(1):30–43.
8. Hongqiang W, Bin D, Yuliang Q. Terahertz radar technology. J Radars. 2018;7(1):1–21.

Chapter 2
Composition and Principles of THz-TDS

The THz-TDS system is based on THz-TDS technology, designed to measure and analyze the spectrum of materials in the terahertz band. A complete THz-TDS system mainly consists of a terahertz source, a terahertz detection device, and a time delay control system. Depending on the measurement mode, the time-domain spectroscopy system can be divided into transmission mode, reflection mode, and attenuated total reflection (ATR) mode. This chapter will introduce the components of the THz-TDS system and the working principles of the measurement modes [1].

2.1 Terahertz Source

The terahertz source is fundamental to the research and application of terahertz technology and is also a significant factor limiting its development. Currently, significant achievements have been made in terahertz radiation sources based on photonics and electronics principles. Photonics sources include terahertz gas lasers, broadband terahertz radiation sources generated by ultrashort pulses, and terahertz radiation sources based on nonlinear optical frequency conversion technology. Electronics sources include terahertz vacuum devices, free-electron lasers, backward wave oscillators, and quantum cascade lasers. In THz-TDS systems, terahertz sources based on photoconductive antennas and optical rectification methods are primarily used.

2.1.1 Photoconductive Antenna

Photoconductive antenna is made by depositing metal electrodes on the surface of a photoconductive semiconductor material to form a dipole antenna structure. It uses a bias voltage applied between the two electrodes to drive free carriers excited

T. Chen et al., *Terahertz Time Domain Spectroscopy*,
https://doi.org/10.1007/978-981-95-2605-5_2

by ultrafast lasers to radiate the pulsed terahertz waves outward. The photoconductive material used for the antenna substrate requires short carrier lifetime, high carrier mobility, and high dark resistivity. Common materials include high-resistivity gallium arsenide (GaAs), indium phosphide (InP), and silicon (Si) wafers with defect structures, with low-temperature grown gallium arsenide (LT-GaAs) being the most used. Antenna structures usually include Hertz dipole antenna structures, resonant dipole antenna structures, and conical antenna structures.

When a photoconductive antenna is used as a terahertz wave emission source, a femtosecond laser pulse is continuously focused on the gap region between the two electrodes of the photoconductive antenna. Since the laser photon energy is higher than the semiconductor material's bandgap, large number of free electron–hole pairs are excited in this region. These electron–hole pairs accelerate under the combined effect of the bias electric field applied to the antenna's two electrodes and the built-in electric field of the semiconductor, forming a transient photocurrent on the semiconductor surface, thereby it radiates terahertz waves outward.

The transient photocurrent density generated in the photoconductive antenna $J_{tx}(t)$ can be expressed as the convolution of the laser pump intensity $P_{opt}(t)$ and the photoconductive antenna pulse response $qn_{tx}(t)v_{tx}(t)$:

$$J_{tx}(t) = P_{opt}(t) * \left[qn_{tx}(t)v_{tx}(t)\right] \tag{2.1}$$

where, $n_{tx}(t)$ represents the photogenerated carrier density, which is a function of carrier lifetime; $v_{tx}(t)$ represents the photogenerated carrier mobility rate, which is related to the applied bias voltage; q is the electron charge. Using the Hertz model for analysis, the far-field terahertz radiation electric field $E_{THz}(t)$ is proportional to the instantaneous photocurrent density in the photoconductive antenna:

$$E_{THz}(t) \propto \frac{dJ_{tx}(t)}{dt} \tag{2.2}$$

The polarization direction of the emitted terahertz wave is determined by the direction of the applied DC bias electric field.

The energy of the terahertz wave radiated by the photoconductive antenna is related to the increase in photocurrent, primarily derived from the potential energy applied to the photogenerated carriers by the external bias electric field. Additionally, a significant portion of the generated terahertz wave energy couples into the substrate material rather than radiating out. Thus, a hyper-hemispherical high-resistivity silicon lens is used in practical applications to enhance the coupling efficiency of terahertz wave radiation into free space. Due to reflection losses at the interfaces between the substrate material and the silicon lens, and between the silicon lens and air, the actual terahertz wave energy emitted by the photoconductive antenna is approximately 15% of the total radiated energy.

2.1.2 Optical Rectification Method

The optical rectification effect is another widely used method for generating terahertz waves. It utilizes the nonlinear effect of a nonlinear crystal under pulsed laser irradiation to produce terahertz waves. Optical rectification refers to the process where incident pulsed laser interacts with a nonlinear medium to generate low-frequency polarization electric field. When laser pulse with a sub-picosecond pulse width is incident on a nonlinear crystal, a low-frequency alternating polarization electric field is generated inside the crystal due to the difference frequency oscillation effect. As shown in Fig. 2.1, nonlinear media commonly used for optical rectification to generate terahertz waves include lithium niobate ($LiNbO_3$) crystals, DAST organic crystals, and semiconductor materials such as gallium arsenide (GaAs) and zinc telluride (ZnTe).

When an ultrafast laser pulse irradiates a nonlinear crystal, the crystal generates a second-order polarization under the influence of the incident laser. Under phase-matching conditions, an alternating polarization electric field related to the envelope of the incident light pulse is formed and radiates terahertz waves. Optical rectification is essentially second-order nonlinear mixing process. Thus, the second-order polarization intensity of the crystal can be expressed as a function of the incident laser field intensity:

$$P^{(2)}(t) \propto \chi^{(2)} E(t)^2 \tag{2.3}$$

where, $\chi^{(2)}$ is the second-order nonlinear susceptibility of the crystal, which is related to the electro-optic coefficient of the crystal. Assuming the incident light field contains two frequency components:

$$E(t) = E_1 e^{-j\omega_1 t} + E_2 e^{-j\omega_2 t} + c.c. \tag{2.4}$$

The second-order polarization of the crystal can be expressed as:

$$\begin{aligned} P^{(2)}(t) \propto \chi^{(2)} \big[& E_1^2 e^{-2j\omega_1 t} + E_2^2 e^{-2j\omega_2 t} + 2E_1 E_2 e^{-j(\omega_1+\omega_2)t} \\ & + 2E_1 E_2^* e^{-j(\omega_1+\omega_2)t} + c.c. \big] + 2\chi^{(2)} \big[E_1 E_1^* + E_2 E_2^* \big] \end{aligned} \tag{2.5}$$

where, $c.c.$ represents the complex conjugate of all terms. The polarization frequency components contained in the equation are the second harmonic, sum frequency,

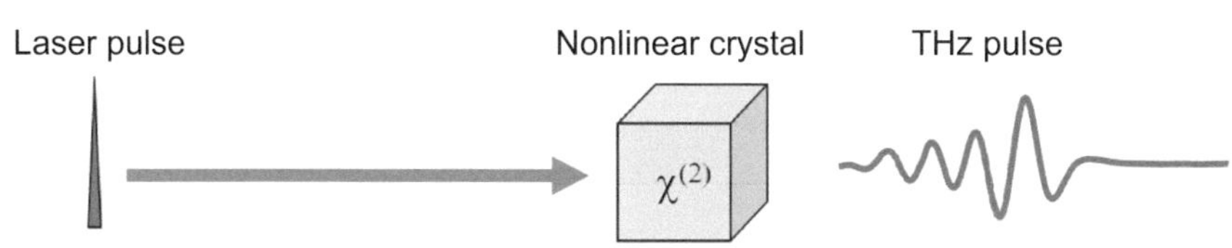

Fig. 2.1 Schematic of terahertz wave generation by optical rectification

difference frequency, and DC components of E_1 and E_2. Among these, only the difference frequency component is related to optical rectification. Since an ultrafast laser pulse contains many frequency components, the polarization oscillation frequency generated by optical rectification can cover a bandwidth ranging from DC to terahertz frequencies, and the intensity of the radiated terahertz wave is proportional to the second derivative of the crystal's second-order polarization intensity:

$$E_{THz}(t) \propto \frac{d^2P^{(2)}(t)}{dt^2} \tag{2.6}$$

Since the polarization intensity of the crystal is determined by the intensity of the incident laser pulse, the intensity of the radiated terahertz wave is also proportional to the second derivative of the laser pulse envelope. The intensity of the incident laser pulse determines the radiation energy of the terahertz wave. The electro-optic coefficient and refractive index of the crystal, as well as the beam waist radius of the excitation laser beam and the duration of the laser pulse, also affects the amplitude and frequency of the radiated terahertz wave.

Since a titanium-sapphire femtosecond laser with a commonly used central wavelength of 800 nm is used as the pump light, zinc telluride crystals can well satisfy the phase-matching conditions. Taking <110> zinc telluride crystals as an example, the relationship between the polarization angle of the pump laser and the terahertz wave radiation is explained. Zinc telluride crystals belong to the zinc blende structure, and their second-order nonlinear coefficient tensor can be expressed as:

$$d = \begin{bmatrix} 0 & 0 & 0 & d_{14} & 0 & 0 \\ 0 & 0 & 0 & 0 & d_{14} & 0 \\ 0 & 0 & 0 & 0 & 0 & d_{14} \end{bmatrix} \tag{2.7}$$

When a linearly polarized light field is incident on the crystal surface in the polarization direction shown in Fig. 2.2.

The nonlinear polarization intensity vector generated by the optical rectification process can be expressed as:

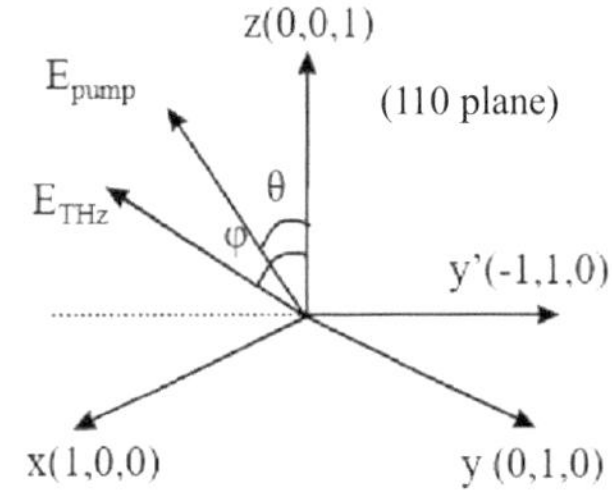

Fig. 2.2 Schematic of terahertz wave radiation coordinates by zinc telluride crystal

$$\begin{bmatrix} P_x \\ P_y \\ P_z \end{bmatrix} = \begin{bmatrix} 0 & 0 & 0 & d_{14} & 0 & 0 \\ 0 & 0 & 0 & 0 & d_{14} & 0 \\ 0 & 0 & 0 & 0 & 0 & d_{14} \end{bmatrix} \begin{bmatrix} E_x^2 \\ E_y^2 \\ E_z^2 \\ 2E_yE_z \\ 2E_xE_z \\ 2E_xE_y \end{bmatrix} \tag{2.8}$$

Here, x, y, z represents the spatial coordinate axes of the crystal lattice, which correspond to the [100], [010], and [001] crystallographic directions. If the incident light field is expressed as:

$$E_{THz} = E_0\left(\frac{\sin\theta}{\sqrt{2}, -\sqrt{2}}\frac{\sin\theta}{, \cos\theta}\right) \tag{2.9}$$

when perpendicular incidents occur, the amplitude and polarization direction of the terahertz radiation electric field generated in the crystal are given by:

$$|E_{THz}| \propto d_{14}E_0^2\left[\sin^2\theta\left(1 + 3\cos^2\theta\right)\right]^{\frac{1}{2}} \tag{2.10}$$

$$\Phi = \arctan(2\cot\theta) \tag{2.11}$$

When $\sin^2\theta = \frac{2}{3}$, i.e., $\theta \approx \pm 55°$, the maximum terahertz wave energy is obtained:

$$E_{THz\,\max} \propto \sqrt{\frac{4}{3}}d_{14}E_0^2 \tag{2.12}$$

2.2 Terahertz Detection

Terahertz wave detection is another key technology in terahertz technology, especially in spectroscopy and imaging technology. Depending on the detection information, it can be divided into coherent and incoherent detection. Coherent detection can simultaneously detect the intensity and phase information of the sample, which includes photoconductive antenna sampling, electro-optic sampling, and air detection. While incoherent detection can only obtain the intensity information of the sample, which includes thermal effect detection, Schottky diodes, and quantum well detectors. In THz-TDS systems, photoconductive antennas and electro-optic sampling methods are commonly used.

2.2.1 Photoconductive Antenna

The same as generation of terahertz waves by a photoconductive antenna, a photoconductive antenna can also be used to detect terahertz waves. The detection process can be regarded as the inverse process of generating terahertz waves by a photoconductive antenna. The detection laser excites electron–hole pairs on the semiconductor surface in the photoconductive antenna region. The terahertz wave converging on the photoconductive antenna provides a bias voltage for the photogenerated carriers, accelerating their movement and forming an instantaneous photocurrent on the semiconductor surface. The intensity of the photocurrent is proportional to the instantaneous electric field of the incident terahertz wave. By measuring the amplified photocurrent intensity, the incident terahertz wave signal can be obtained.

Same as terahertz generation device, photoconductive antenna can be used to detect terahertz wave. But its working process is reversed. As shown in Fig. 2.3. The slit region of the photoconductive antenna is excited by the detection light to ensure a fixed relationship between the terahertz pulse and the detection light pulse. When the terahertz wave and the detection light simultaneously illuminate the slit region between the two electrodes of the photoconductive antenna in the same polarization direction. The incident terahertz wave on the semiconductor surface formed alternating electric field, which is simultaneously excited by the detection laser, thus the free photogenerated carriers generates a transient photocurrent:

$$J_{rx}(t;\,\Delta t) = P_{opt}(t) * \left[qn_{rx}(t)v_{rx}(t;\,\Delta t)\right] \tag{2.13}$$

Here, the photocurrent density of the detection antenna $J_{rx}(t;\,\Delta t)$ and the average velocity of the carriers $v_{rx}(t;\,\Delta t)$ are the functions of delay time between the pump light and the detection light in the system, which is determined by the received terahertz wave electric field $E_{THz}(t + \Delta t)$. Therefore, the actual measured current is:

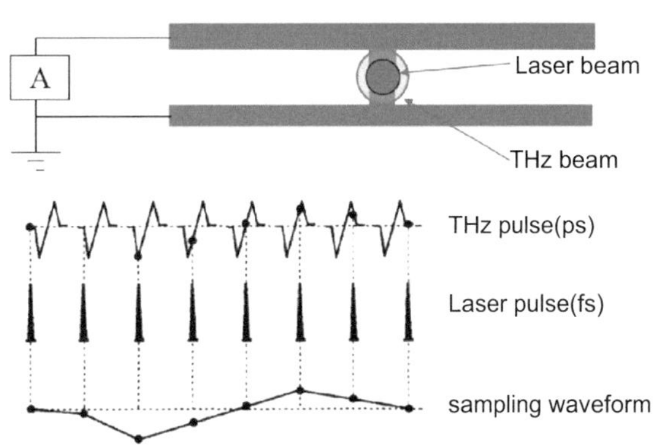

Fig. 2.3 Schematic of terahertz wave detection by photoconductive antenna

$$\bar{J}_{rx}(\Delta t) \propto \int_{-\infty}^{+\infty} J_{rx}(t; \Delta t)dt \tag{2.14}$$

2.2.2 Electro-Optic Sampling

The process of electro-optic sampling detection of terahertz waves using a nonlinear crystal can be regarded as the inverse process of optical rectification. The polarization state is shown in Fig. 2.4. When linearly polarized detection light and terahertz waves are collinearly focused on the crystal, due to the electro-optic effect (Pockels effect), the terahertz wave electric field changes the birefringence properties of the crystal, and the incident linearly polarized detection light becomes elliptically polarized. After passing through a quarter-wave plate and a Wollaston prism, the intensity difference between the two orthogonal components of the detection light is proportional to the intensity of the incident terahertz electric field. By changing the time relationship between the detection light and the terahertz wave, the amplitude and phase information of the terahertz wave electric field can be obtained based on the received photocurrent signal.

Taking a <110> zinc telluride crystal as an example, its coordinate schematic is shown in Fig. 2.5. When a terahertz wave is incident perpendicularly on the crystal, the refractive index ellipsoid equation of the crystal becomes:

$$\frac{x^2 + y^2 + z^2}{n_0^2} + \frac{2\gamma_{41} E_{THz} \sin\alpha}{\sqrt{2}} yz - \frac{2\gamma_{41} E_{THz} \sin\alpha}{\sqrt{2}} zx + 2\gamma_{41} E_{THz} \cos\alpha xy = 1 \tag{2.15}$$

After two coordinate transformations and simplifications, the received photocurrent signal is calculated as

$$\Delta I(\alpha, \varphi) = I_p \frac{\omega n^2 E_{THz} \gamma_{41} L}{2c} (\cos\alpha \sin 2\varphi + 2 \sin\alpha \cos 2\varphi) \tag{2.16}$$

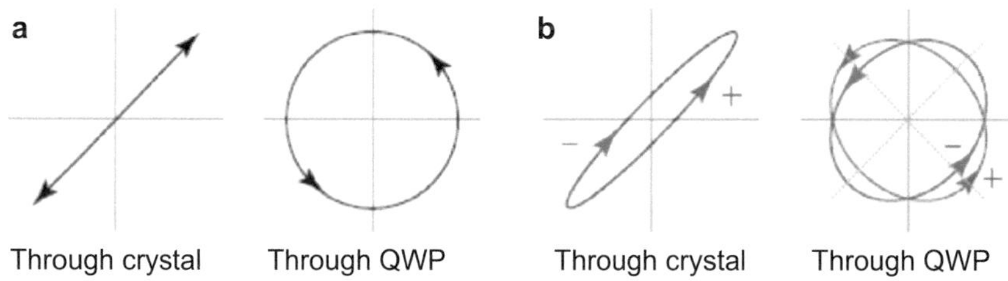

Fig. 2.4 Detection of light polarization state: **a** without terahertz wave modulation; **b** with terahertz wave modulation

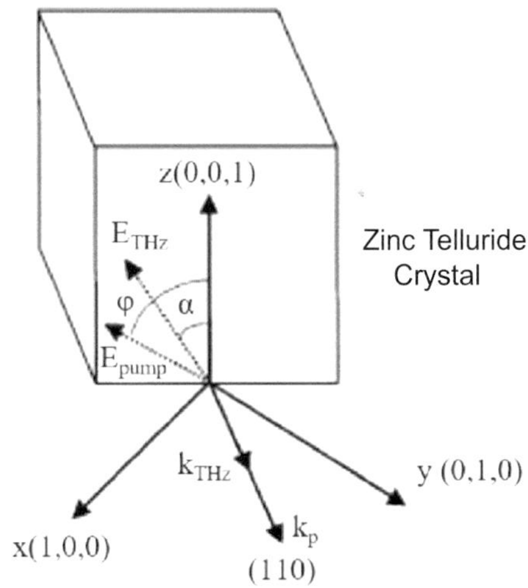

Fig. 2.5 Schematic of terahertz wave detection coordinates by zinc telluride crystal

where, I_p is the detected light intensity, L is the crystal thickness, γ_{41} is the linear electro-optic coefficient of the crystal, c is the speed of light in vacuum, ω is the angular frequency of the detected light. The received photocurrent signal ΔI is proportional to the terahertz wave electric field E_{THz} and is related to α, the angle between the polarization direction of the terahertz wave and the crystal axis, as well as φ, the angle between polarization direction of the detected light and the crystal axis.

2.3 Delay Control Methods

In a THz-TDS system, the terahertz wave signal excited by a femtosecond laser is a pulse signal with a repetition rate consistent with the pump femtosecond laser. For this type of signal, there are usually two sampling methods: real-time sampling and equivalent time sampling. The principles of these two sampling methods are shown in Fig. 2.6. Real-time sampling refers to extracting evenly spaced sampling points within a certain time to sufficiently reproduce the original signal. The Nyquist sampling theorem must be satisfied, which means that the repetition rate of the detected light must be at least twice that of the pump light. Equivalent time sampling is for periodic signals, where signal points at different positions in different periods are extracted and then the signal is reconstructed. For THz-TDS, an ultrafast electro-optic sampling system, real-time sampling is difficult to achieve, while equivalent time sampling is not constrained by the Nyquist sampling theorem and is suitable for sampling periodic signals.

The key to achieving equivalent time sampling of terahertz radiation lies in the time delay between the femtosecond lasers at the terahertz emitter and detector

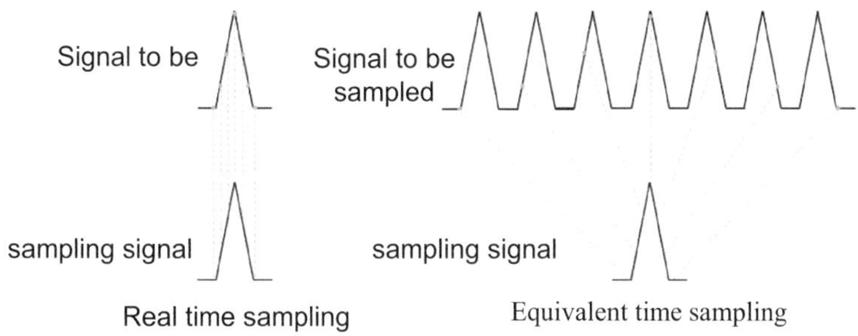

Fig. 2.6 Schematic of real-time sampling and equivalent time sampling principles signal to be sampled

side. Currently, there are two methods to achieve delay: mechanical delay and asynchronous delay.

2.3.1 Mechanical Delay

The hardware implementation of mechanical delay involves placing two right-angle mirrors on a precision moving stage. As the detected light passes through these mirrors, the continuous movement of the stage changes its optical path, creating a time delay. Assuming the platform moves in the direction of increasing optical path, the time delay is:

$$\Delta t = t_2 - t_1 = \frac{2n\Delta L}{c} \tag{2.17}$$

where n is the refractive index, and c is the speed of light. By continuously varying the optical path difference between the pump light and the detected light, i.e., the time delay, equivalent time sampling of different sampling points on continuous periods of the terahertz periodic signal is achieved. Although this mechanical delay method is simple and easy to operate, it requires the mechanical moving stage to be strictly collinear and highly stable. Additionally, due to the limited speed of the mechanical stage, the system's time resolution is relatively low, and the sampling time is usually long.

2.3.2 Asynchronous Sampling Delay

Asynchronous sampling delay technology is a delay control method that achieves equivalent time sampling without relying on any mechanical moving devices. It uses two femtosecond lasers with slightly different repetition rates as the pump light and the detected light, replacing the process with the mechanical delay method. It assumes the terahertz pulse excited by the pump light coincides with the detected light pulse in a certain period. Due to the different repetition rates, a time delay is added with each pulse until the two pulses coincide again in a certain period, which completes the full sampling of a terahertz pulse signal.

Assuming the number of sampling points is n, and the repetition rates of the pump light and the detected light are f_b and f_p, respectively, where $f_b = f_p + \Delta f$, then the fixed time delay between the two femtosecond laser pulses is:

$$\Delta t = \frac{1}{f_p} - \frac{1}{f_b} \tag{2.18}$$

The relationship between the sampling points and the repetition rates of the two beams of light is as follows:

$$\frac{n}{f_b} = \frac{n-1}{f_p} \tag{2.19}$$

Therefore, the time required to detect a complete terahertz pulse is:

$$T = \frac{1}{\Delta f} \tag{2.20}$$

Typically, THz-TDS system based on asynchronous sampling delay can acquire a complete terahertz signal in just milliseconds, whereas mechanical delay requires several tens of seconds. Therefore, THz-TDS systems utilizing asynchronous sampling delay technology have significant advantages in detection time and resolution.

2.4 Measurement Modes

To meet the testing needs for different types of samples, THz-TDS systems can be primarily divided into transmission mode, reflection mode, and attenuated total reflection mode. Among these, transmission mode is the most widely used measurement mode, which is applicable to solid non-polar samples, liquid thin layers, gas chambers, etc. Reflection mode is mainly for samples that cannot be penetrated, such as very thick samples or metallic samples. Attenuated total reflection uses evanescent waves to measure the interface between the sample and a silicon prism, requiring close contact, and is generally used for measuring liquid and powder samples.

2.4.1 Transmission Mode

The transmission mode of THz-TDS systems is most widely applied. Taking system using a photoconductive antenna as the emitter and detector as an example, in THz-TDS system, the femtosecond laser beam emitted by the laser is divided into pump light and probe light, which travel along the pump path and probe path, respectively. At the terahertz wave emission end, the pump light is focused by an optical lens onto a photoconductive antenna or nonlinear crystal along the pump path, generating terahertz waves. A pair of off-axis parabolic mirrors are then used to collimate and focus the emitted terahertz waves onto the sample. After passing through the sample, the terahertz waves carrying information about the sample's properties are collimated and focused by another pair of off-axis parabolic mirrors and received by the detector. At the terahertz wave detection end, the probe light is focused onto the detector by an optical lens after passing through a delay path. The detector outputs a current signal related to the terahertz signal, which is then amplified by a lock-in amplifier to obtain the final measurement signal, and a Fourier transform is performed to obtain the corresponding terahertz spectrum.

2.4.2 Reflection Mode

The reflection mode THz-TDS system is designed to measure materials that strongly reflect or partially reflect terahertz waves. According to Fresnel's law, the reflectance is related to the complex refractive index of the sample material, so the material parameters of the sample can also be extracted from the reflection signal. Compared to transmission-type THz-TDS systems, the detection part of reflection-type systems is not placed behind the sample. In the case of oblique incidence reflection measurements, the terahertz detector is positioned at a certain angle to the direction of the incident terahertz wave; for normal incidence reflection mode, a beam splitter placed in the incident light path is used to reflect the detection signal to the detector. Generally, reflection-type THz-TDS systems can be divided into two main categories. One is the single reflection measurement system, where the incident terahertz wave reflects once on the sample surface and is directly received by the terahertz wave detection element.

Another type is the double reflection path THz-TDS measurement system, also known as a self-referencing structure, which requires processing a high-resistance silicon transmission window structure on the sample surface. During measurement, the signal received by the detector includes the first reflection signal of the terahertz wave at the interface between the window material and air, and the second reflection signal at the interface between the window material and the sample material. These two detected reflection signals are used as reference and sample signals to calculate the sample parameters.

2.4.3 Attenuated Total Reflection Mode

The terahertz attenuated total reflection detection mode combines the advantages of terahertz reflection detection and attenuated total reflection. It enables the detection of samples with significant thickness or strong terahertz absorption, such as heavily doped semiconductor carriers and polar liquid molecules. Compared to the reflection-type system, an attenuated total reflection prism, usually a silicon prism, is added beneath the sample.

The principle of attenuated total reflection is illustrated in Fig. 2.7. The reflection and refraction of terahertz waves at the interface can be described by Snell's law. When incident light travels from an optically dense medium to an optically rarer medium, the refraction angle is greater than the incidence angle. When the refraction angle equals 90°, the incidence angle becomes the critical angle. When the incidence angle is less than the critical angle, ordinary reflection occurs; when it is greater than the critical angle, total reflection occurs. During total reflection, part of the incident terahertz wave enters the sample as an evanescent wave, whose intensity decays exponentially in the penetration direction and propagates along the interface. Consequently, there is a very short lateral displacement on the reflection interface, known as the Goos-Hänchen shift, meaning the incident and reflection points are not at the same location. Since the sample is not transparent to the terahertz wave entering as an evanescent wave, phenomena such as absorption and dispersion occur. Thus, the reflected wave becomes weaker than the incident wave and carries the optical information of the sample. This is referred to as attenuated total reflection.

Evanescent waves are a type of surface wave suitable for probing the shallow regions of a sample. Therefore, compared to reflection and transmission modes, attenuated total reflection does not impose requirements on the size and morphology of the sample. Even if the morphology is irregular, interaction with the sample can occur in the form of evanescent waves to extract the sample's optical parameters. Additionally, the attenuated total reflection signal is highly sensitive to the sample's dispersion relationship and offers high sensitivity.

Fig. 2.7 Schematic of attenuated total reflection

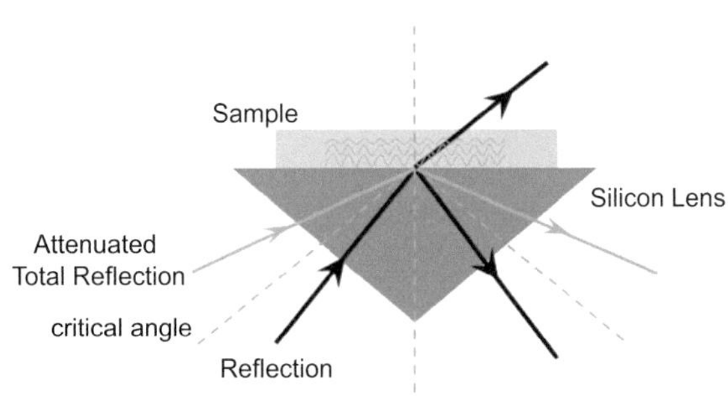

Reference

1. Jining L. Research on mechanically tunable terahertz metamaterials. Tianjin: Nankai University; 2013.

Chapter 3
THz-TDS Data Processing Methods

The previous chapter introduced the hardware components of the THz-TDS system, including the pump femtosecond laser source, terahertz generation source, terahertz detector, and delay controller. Here, a complete algorithm is provided for converting the collected terahertz time-domain pulse signals into terahertz spectral data of the sample. This chapter will introduce the basic physical parameter extraction methods for transmission and reflection THz-TDS techniques, including the extraction of optical coefficients from terahertz time-domain transmission spectra based on the principle of multiple beam interference. Finally, the core measurement parameters of THz-TDS are introduced.

3.1 Extraction of Physical Parameters in THz-TDS

The theoretical calculation model for extracting optical parameters of samples using THz-TDS was first proposed by Duvillaret et al. [1], including different calculation principles for transmission and reflection THz-TDS systems. The principles for extracting terahertz spectral parameters are introduced below [2].

3.1.1 Extraction of Transmission THz-TDS Parameters

First, the measurement of the transmission THz-TDS system is introduced, as shown in Fig. 3.1. The transmitted waveforms with and without the sample are displayed, where $E_{in}(t)$ represents the incident terahertz pulse, $E_R(t)$ represents the reference pulse without the sample, and $E_S(t)$ represents the terahertz pulse transmitted through the sample.

T. Chen et al., *Terahertz Time Domain Spectroscopy*,
https://doi.org/10.1007/978-981-95-2605-5_3

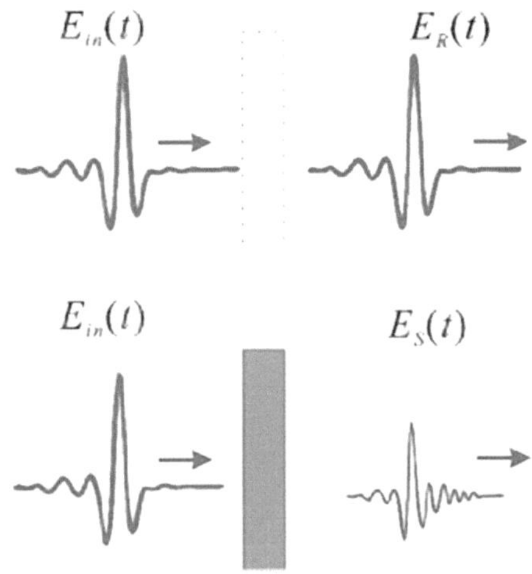

Fig. 3.1 Schematic of terahertz wave electric field scanning with and without sample

By detecting the interaction between the terahertz background pulse and the terahertz pulse transmitted through the sample using a terahertz detector, we can obtain the transmission electric field signal in the time domain. The corresponding frequency domain spectrum signal is obtained through Fourier transform. When the terahertz pulse passes through the sample, the transmission spectra with and without the sample should be

$$E_R(\omega) = E_{in}(\omega)\exp(ik_R d) \tag{3.1}$$

$$E_S(\omega) = E_{in}(\omega)\frac{t_{as}t_{sa}\exp(ik_S d)}{1 + r_{as}r_{sa}\exp(2ik_S d)} \tag{3.2}$$

where, r_{as}, r_{sa}, t_{as}, and t_{sa} are the Fresnel reflection and transmission coefficients at the air-sample and sample-air interfaces, respectively. $E_{in}(\omega)$, $E_R(\omega)$, $E_S(\omega)$ are the spectra of the incident terahertz wave, the background reference terahertz wave without the sample, and the terahertz wave transmitted through the sample, respectively. k_R and k_S are the wave vectors of the terahertz wave, c is the speed of light. Due to multiple reflections on the two surfaces of the sample, a main peak appears in the time-domain signal is followed by several smaller peaks. The main peak and the smaller peaks caused by multiple reflections are very easy to separate, so it is necessary to remove the signals after the main peak and analyze only the main peak. For this simple case, Eq. (3.2) can be simplified as

$$E_S(\omega) = E_{in}(\omega)t_{as}t_{sa}\exp(ik_S d) \tag{3.3}$$

Thus, the ratio of the electric field with and without the sample can be derived as

$$\frac{E_S(\omega)}{E_R(\omega)} = A\exp(i\Phi) = \frac{4n_S(\omega)}{[n_S(\omega)+1]^2}\exp[-\alpha_S(\omega)d/2] \times \exp\{i[n_S(\omega)-1]]\omega d/c\} \tag{3.4}$$

where A and Φ are the amplitude and phase obtained through Fourier transform, respectively, and α_S is the absorption coefficient. By comparing the amplitude and phase of the sample and the reference background, the real part of the complex refractive index, the refractive index n_S, and the imaginary part, the extinction coefficient $\kappa_s(\omega)$, can be determined separately.

$$n_S(\omega) = 1 + \frac{\Phi c}{\omega d} \tag{3.5}$$

$$\kappa_s(\omega) = -\frac{c}{\omega d}\ln\left[\frac{[n_S(\omega)+1]^2}{4n_S(\omega)}A\right] \tag{3.6}$$

Thus, the absorption coefficient α_S can also be obtained.

$$\alpha_S(\omega) = 2\frac{\kappa_s(\omega)\omega}{c} = -\frac{2}{d}\ln\left[\frac{[n_S(\omega)+1]^2}{4n_S(\omega)}A\right] \tag{3.7}$$

From the above description, THz-TDS can be used to study the properties of sample materials in the terahertz frequency range. The following parts will introduce the role of this method in the study of actual sample materials and specifically introduce the measurement results and analysis of several materials.

3.1.2 Reflection Mode THz-TDS Parameter Extraction

The extraction of medium parameters of the measured sample in reflection mode is similar to transmission mode. In the case of normal incidence and a planar reflective surface, the reflected wave can be expressed as:

$$E_{refl} = -E_{inc}\frac{\tilde{n}' - \tilde{n}}{\tilde{n}' + \tilde{n}} \tag{3.8}$$

where $\tilde{n}$ is the negative refractive index of the medium where the incident electromagnetic wave is located before reflection, and $\tilde{n}'$ is the negative refractive index of the medium causing the reflection. By comparing the terahertz wave electric field intensity containing sample information with the reference signal electric field intensity without sample information, we can obtain:

$$\frac{E_s(\nu)}{E_{ref}(\nu)} = |r|e^{i\Phi} = \frac{\tilde{n}-1}{\tilde{n}+1} \tag{3.9}$$

Here, the incident wave is incident from the air, and the reference signal comes from a perfectly reflecting surface for terahertz waves. In practical experimental systems, a gold or silver-coated planar mirror is generally used to approximate this. Substituting the negative refractive index $|\tilde{n} = n + i\kappa|$ into the above equation, it can be obtained:

$$|r|e^{i\Phi} = \frac{n + i\kappa - 1}{n + i\kappa + 1} \tag{3.10}$$

From this, the refractive index and absorption coefficient of the sample can be obtained as:

$$n = \frac{1 - |r|^2}{1 + |r|^2 - 2|r|\cos\Phi} \tag{3.11}$$

$$\alpha = \frac{4\pi\nu}{c} \cdot \frac{2|r|\sin\Phi}{1 + |r|^2 - 2|r|\cos\Phi} \tag{3.12}$$

It is worth noting that in reflection mode measurements, since the reference reflection signal is provided by the mirror, any error in the position of the reference mirror and the sample will strongly affect the phase change of the signal in the measurement results. Thereby, it will affect the calculation and extraction of the refractive index and absorption coefficient.

From the expression of the absorption coefficient of the medium above, it can be found that when $r = 1$ and $\phi = 0$, the absorption coefficient reaches its maximum value, at which point:

$$\alpha \approx \frac{4\pi\nu}{c} \cdot \frac{2\Phi}{(1 - |r|)^2 + \Phi^2} \tag{3.13}$$

In actual reflection measurement experiments, if the reflection coefficient is $|r| = 1 \pm s_r$ and the phase is $\phi = 0 \pm s_\phi$, the absorption coefficient is:

$$\alpha \frac{4\pi\nu}{c} \frac{2s_\Phi}{s_r^2 + s_\Phi^2}{}_{\max} \tag{3.14}$$

3.1.3 *Terahertz Time-Domain Transmission Spectrum Optical Coefficient Extraction Based on Multi-Beam Interference Principle*

When characterizing the optical coefficients of thin parallel plane samples using a THz-TDS system, there is a problem of periodic Fabry-Pérot (F-P) oscillations caused by multiple reflections, as shown by the weaker pulse signals in Fig. 3.2. These

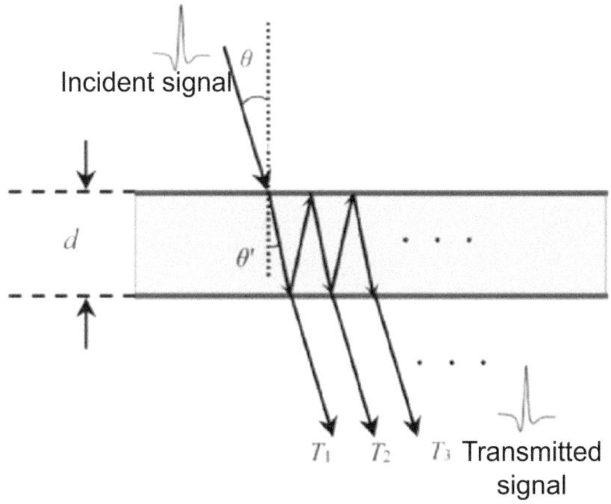

Fig. 3.2 Multi-layer interference optical path in parallel plate sample

multiple reflections cause spectral modulation in the frequency domain. A common method is to remove the small peaks of the time-domain signal generated by multiple reflections, but since the removed signal may contain useful sample information, data processing may be inaccurate. Although complex numerical analysis algorithms have been proposed in early reports, the accuracy of existing algorithms is limited by the uncertainty of the measurement signal. The echo pulses caused by the Fabry-Pérot effect can be used to determine the sample thickness and optical coefficients, but the iterative process and modeling of the transfer function fitting are very complex [3–8].

The following part introduces a simple and effective method for extracting optical coefficients based on typical THz-TDS system hardware. By utilizing the complete time-domain spectrum without removing echo pulses, the oscillating transmission spectrum in the frequency domain is obtained. The refractive index and absorption coefficient are derived based on the principles of multiple beam interference and the Fresnel equations. This method differs from traditional methods by processing data independently of phase. Finally, experimental results using this method on polyethylene (PE) and quartz samples of different thicknesses are provided, it shows good agreement with traditional data processing methods and verifies the reliability of this approach.

In THz-TDS transmission measurement mode, the transmittance of normally incident sample can be given by the Fresnel equations, where the transmission coefficient T is determined by the real part $n(\nu)$ and the imaginary part $\kappa(\nu)$ of the complex refractive index. It can be expressed as

$$T = \frac{2\sqrt{n}}{n+1} \tag{3.15}$$

$$n = n(\nu) + i\kappa(\nu) \tag{3.16}$$

The transmission coefficient T includes the amplitude and phase of the electric field. The transmittance $T(\nu)$ is related to the transmission coefficient, it can be expressed as follows:

$$T(\nu) = |T|^2 = \frac{I_T}{I_0} \tag{3.17}$$

where I_T and I_0 are the transmitted intensity and reference intensity, respectively.

For thin film samples, Fig. 3.2 shows the general case with an incident angle of θ and a refraction angle of θ', where I_{T1}, I_{T2}, and I_{T3} correspond to 0, 1, and 2 round-trip delays of transmitted light intensity, respectively. The interference intensity at a specific wavelength is determined by the optical path difference (OPD) between consecutive transmitted beams, it can be expressed as:

$$OPD = 2nd\cos\theta' \tag{3.18}$$

where d is the sample thickness and n is the real part of the refractive index. Under the approximation of near-normal incidence θ, the OPD can be simplified to:

$$OPD = 2nd \tag{3.19}$$

For the broadband terahertz source used in the THz-TDS system, oscillation peaks appear on the frequency domain transmission curve. Constructive interference occurs when the phase difference is an integer multiple of 2π or the OPD is a multiple of the wavelength λ. Assuming any maximum intensity appears at interference order m at λ_1, the adjacent maximum is at interference order $m + 1$ at λ_2 ($\lambda_2 < \lambda_1$). The relationship between wavelength and interference order is given by:

$$2n_1 d = m\lambda_1 \tag{3.20}$$

$$2n_2 d = (m + 1)\lambda_2 \tag{3.21}$$

Assuming the refractive index does not change rapidly, due to the small frequency interval, it can be assumed that the refractive indices n_1 and n_2 are equal. The equation can be simplified to:

$$2nd = m\lambda_1 = (m + 1)\lambda_2 \tag{3.22}$$

This equation can be used to obtain the interference order m and refractive index n at λ_1 (or λ_2). This process can be extended to all frequencies to obtain the refractive index over the measurement range. No preprocessing of the original data is performed in this process to ensure the integrity of the sample information.

The extinction coefficient $\kappa(\nu)$ and the related absorption coefficient α can be calculated based on the Fresnel equations, which combine the complex electric field transmission formula and the transmittance obtained directly from the Fourier transform of the time-domain signal:

$$\kappa(\nu) = -\ln\left\{\frac{[1+n(\nu)]^2}{4n(\nu)}T(\nu)\right\}\frac{c}{2\pi\nu d} \tag{3.23}$$

$$\alpha(\nu) = \frac{4\pi\kappa(\nu)}{\lambda} \tag{3.24}$$

If needed, the complex dielectric parameters, including the real part ε_r and the imaginary part, as functions of the refractive index $n(\nu)$ and the extinction coefficient $\kappa(\nu)$, are described as follows:

$$\varepsilon_r(\nu) = n^2(\nu) - \kappa^2(\nu) \tag{3.25}$$

$$\varepsilon_i(\nu) = 2n(\nu)\kappa(\nu) \tag{3.26}$$

To verify the applicability of the above method, a series of samples were studied using a THz-TDS system, including two materials: white polyethylene and quartz, with thicknesses ranging from 0.30 to 3 mm, and an aperture greater than 30 mm. The quartz samples have a roughness of 7 μm commercial slides, while the polyethylene samples are made from industrial materials and prepared during cutting, with an estimated roughness of 20 μm. The raw data obtained from the THz-TDS system is the time-domain signal of the terahertz electric field. Taking the white polyethylene sample with a thickness of 1.99 mm shown in Fig. 3.3a as an example, due to multiple reflections between two parallel surfaces, two terahertz peaks P_1 and P_2 can be observed in the time-domain spectrum. Due to the round-trip attenuation within the sample, the intensity of the higher-order peak P_2 is lower. For certain semiconductor materials (such as silicon, germanium, gallium arsenide, etc.), the round-trip loss is smaller, and more peaks may be observed. In traditional data processing, only the main peak P_1 is considered useful, and all remaining parts are removed to eliminate oscillations in the frequency domain, which otherwise makes it difficult to obtain accurate absorption or transmission spectra. When peaks are close to each other, it becomes challenging to determine the correct data range. In Fig. 3.3a, two sensitive nodes marked as n_1 and n_2 can be used as cutoff points before Fourier transforming to the frequency domain. As shown in Fig. 3.3b, if the cutoff node (n_1) is close to the main peak P_1, much detailed information will be lost in the transmission spectrum. Considering all information, the differences in the absorption spectrum will be significantly greater. Setting the range at n_2 should be a relatively ideal choice, as most absorption information is included without being affected by the secondary peak P_2. Sometimes, the terahertz field containing absorption information extends to a wider time delay range, making it difficult to choose an appropriate cutoff node for thin samples. During data processing, if the signal acquisition position is further extended to the complete time domain to include information from the secondary peak P_2, interference will begin to appear, and frequency domain oscillations prevent us from

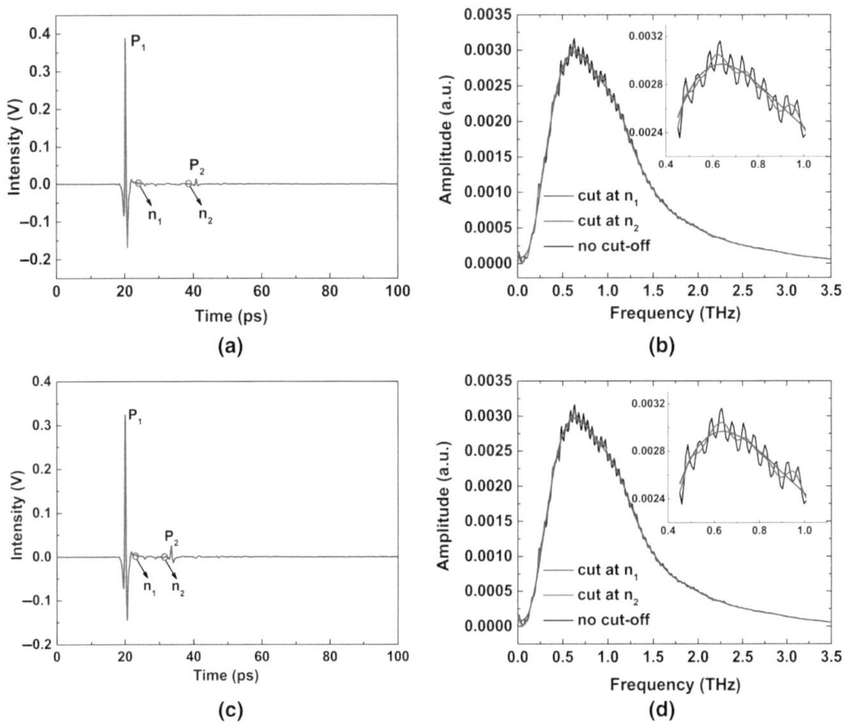

Fig. 3.3 Terahertz time-domain signal of white polyethylene (**a**) and quartz sample (**c**), frequency domain spectrum at different cutoff positions in (**b**) and (**d**)

directly obtaining accurate transmission spectra. Quartz samples face the same issue, and the measurement results also encounter this problem, as shown in the terahertz time-domain and frequency-domain spectra of quartz samples in Fig. 3.3c, d.

For samples with parallel surfaces, multiple reflections within the sample will result in periodic Fabry-Pérot oscillations in the frequency domain. Taking the 1.99 mm white polyethylene sample as an example, under normal incidence, the oscillations in the transmission spectrum are shown in Fig. 3.4.

To retain all sample information, a Fourier transform is applied over the entire time-domain window. According to the above theoretical analysis, the frequency interval Δν marked in Fig. 3.4 is determined by the sample, with thicker samples having smaller Δν values, and the results for refractive index and absorption coefficient should be more accurate. For thicker samples, due to the introduction of more absorption, the amplitude of frequency domain oscillations will decrease, leading to a reduced frequency range for extracting the refractive index due to a lower signal-to-noise ratio. Considering frequency resolution and signal-to-noise ratio, each sample has an optimal thickness [4]. On the other hand, when echoes overlap, it is almost impossible to process data for very thin samples in the order of hundreds of micrometers using traditional methods, but it can be calculated through interference fringes.

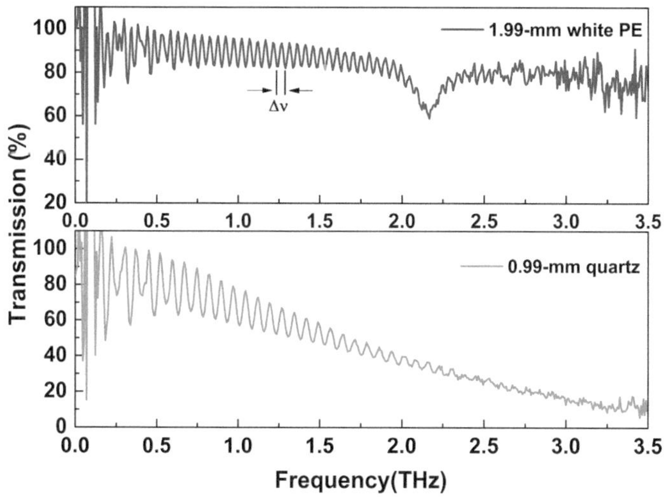

Fig. 3.4 Transmission spectrum of polyethylene and quartz

When the optical path difference is a multiple of the corresponding wavelength, interference fluctuations appear in the transmission spectrum, and the refractive index can be calculated using the above formula.

Figure 3.5 shows the terahertz spectral results for white polyethylene samples with thicknesses of 0.82, 1.99, and 3.01 mm. There is a distinct absorption peak in the absorption spectrum located around 2.16 THz. By using the multi-beam interference method to extract data, all interference orders from 1 to 64 within the effective frequency range are calculated. The deviation of the 0.82 mm sample is caused by larger thickness fluctuations and inaccuracies, which is a common issue for thinner samples. The divergence at the high-frequency and low-frequency ends of the effective range extraction should be attributed to the system's lower signal-to-noise ratio at corresponding frequencies. The results obtained using traditional data processing methods of amplitude and phase spectra are indicated by dashed lines, and all sample data show good consistency, with no significant differences between the two data processing methods, verifying the method's effectiveness. The refractive index and absorption coefficient of three quartz samples with thicknesses of 0.30, 0.53, and 0.99 mm are shown in Fig. 3.5b, d, and the measurement results for quartz samples are also consistent between the two methods.

In transmission mode, the applicability of the multilayer interference method for extracting optical coefficients is almost identical to traditional methods. However, the multilayer interference method requires high-contrast interference fringes, resulting in a slightly narrower effective frequency range. In experiments, the polyethylene samples were not fully polished, and due to impurities, the material distribution was not uniform. Nevertheless, effective echoes and optical coefficients could still be obtained without stringent sample preparation, and slight scattering does not affect the results. However, if scattering and absorption are too severe, the time-domain

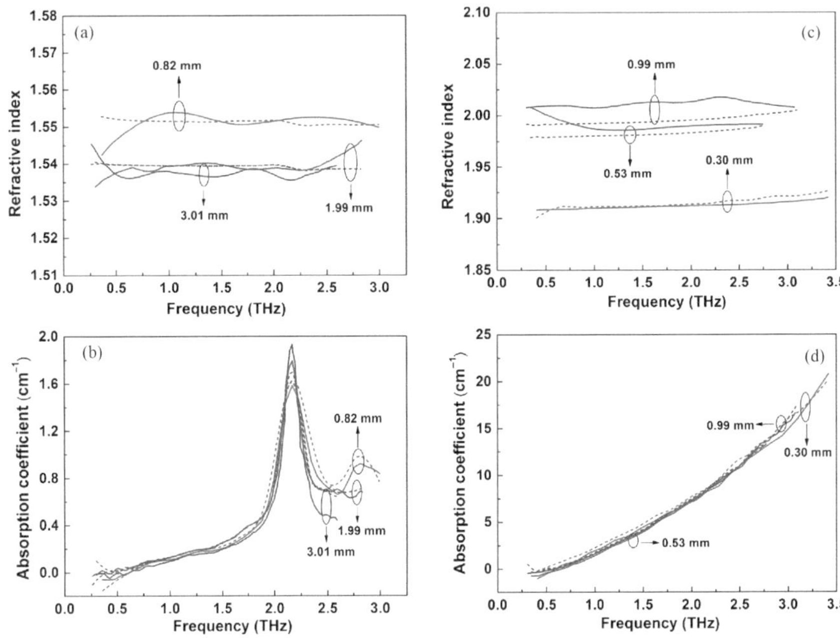

Fig. 3.5 Extraction of optical coefficients using multi-beam interference method (Solid Line) and traditional method (Dashed Line): (**a**) and (**b**) refractive index and absorption coefficient of white polyethylene samples with different thicknesses; (**c**) and (**d**) refractive index and absorption coefficient of quartz samples with different thicknesses

echoes may disappear, rendering this method unsuitable. Using the best-processed 1.99 mm white polyethylene sample and 0.99 mm quartz sample as examples, the relative measurement error of their optical coefficients is summarized. Except for individual frequency points near the limit range, the refractive index error is within 1%, and the absorption coefficient error is below 10%.

Such relative errors are acceptable in most cases when describing the material properties of samples. For thin samples, maintaining good parallelism during sampling and polishing can reduce measurement errors. For a 1.99 mm polyethylene sample below 1 THz, the thickness measurement error is 5 μm, corresponding to a refractive index error of about 0.002. Therefore, the parallelism, polishing, and precise thickness measurement of the sample are crucial for accurately measuring optical coefficients. Additionally, the incident terahertz beam should be perpendicular to the sample surface; otherwise, the effective optical path may change, affecting the precision of the interference order. According to calculations, the error caused by the incident angle is an order of magnitude lower than that caused by sample thickness and it is easier to maintain approximately normal incidence during data acquisition. A stable THz-TDS system is conducive to precise measurements. The larger relative error in the absorption coefficient is likely due to system fluctuations, as absorption calculations rely more on terahertz intensity than on the refractive index. Thicker samples should have better frequency resolution when encountering rapid changes in the frequency domain; accuracy should be better, but if absorption is severe, the signal-to-noise ratio of thicker samples is lower. For thinner samples, the resolution is lower but compared to traditional THz-TDS data processing methods, when time-domain echoes overlap and oscillations in the frequency domain spectrum cannot be eliminated, the multibeam interference method can provide more accurate refractive index and absorption coefficient results. In most THz-TDS systems, samples are not placed in parallel beams but in focused beams. Because of divergence on the power transfer function, an overly simplified plane wave assumption may be inaccurate.

Compared to traditional data processing methods, the multibeam interference method is more affected because multiple sample thicknesses are involved in the raw data. Therefore, to simplify and minimize measurement errors, the sample thickness should be less than the order of 1 cm, and the sample must be located at the center of the terahertz beam waist [6].

3.2 Key Indicators of THz-TDS

The core indicators for measuring the performance of THz-TDS include spectral range, dynamic range, spectral resolution, etc.

3.2.1 Spectral Range

The spectral range refers to the upper and lower limits of the spectrum that a THz-TDS system can effectively detect. Typically, when the signal-to-noise ratio drops to 1, it is defined as the nominal spectral range of the system, while the practical spectral range and high-quality spectral range are usually considered to be regions where the signal is greater than −40 dB and −20 dB, respectively.

The spectral range of a THz-TDS system is determined by the terahertz radiation source used. Currently, the terahertz radiation sources in commercial systems mainly include photoconductive antenna radiation sources, Cherenkov sources based on lithium niobate crystals, and broadband radiation sources based on organic crystals. The spectral range of photoconductive antenna radiation sources is relatively small, mostly concentrated between 0.1 and 4 THz due to differences in the femtosecond lasers used. Figure 3.6 shows the spectral range of a terahertz source based on a photoconductive antenna. Cherenkov radiation sources based on lithium niobate crystals can be broadened to around 7 THz; radiation sources based on organic crystals are usually broadband, with a spectral range from about 0.3–15 THz. Considering that the introduction of samples will attenuate the terahertz signal intensity, the applicable range of a THz-TDS system is smaller than the nominal spectral range.

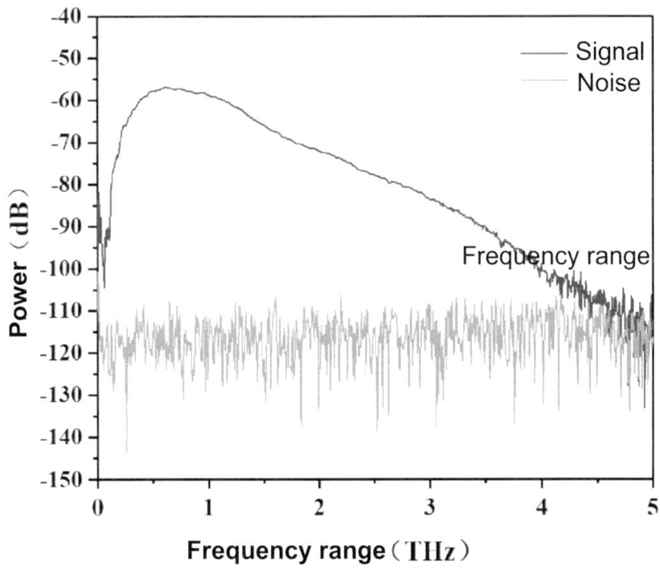

Fig. 3.6 Power spectrum of THz-TDS

3.2.2 *Dynamic Range*

Due to the relatively low power of THz-TDS systems, the dynamic range of the terahertz signal is generally used to measure the system's signal-to-noise ratio. The dynamic range refers to the difference between the maximum and minimum power of the terahertz signal, with the minimum value typically being the intensity of the noise signal. As shown in Fig. 3.6, the noise baseline is at −120 dB, and the maximum value of the terahertz signal is −60 dB, resulting in a system dynamic range of 60 dB. Currently, mature THz-TDS systems in use generally achieve a dynamic range of 60–70 dB or more, while high-performance systems can reach a dynamic range of over 100 dB.

3.2.3 *Spectral Resolution*

Spectral resolution refers to the minimum time or wavelength interval between sampling points that the system can distinguish, corresponding to time resolution and frequency resolution. Generally, the spectral resolution of a THz-TDS system is determined by factors such as the repetition rate of the femtosecond laser, time delay, and the number of sampling points. Typically, higher resolution implies a longer sampling time. In THz-TDS testing, a spectral resolution of less than 5 GHz is generally considered as high resolution, while a resolution of 10–20 GHz is considered standard. The highest spectral resolution of THz-TDS systems has reached below 1 GHz.

References

1. Duvillaret L, Garet F, Coutaz JL. Highly precise determination of optical constants and sample thickness in THz-TDS. Appl Opt. 1999;38(2):409–15.
2. Jining L. Research on mechanically tunable terahertz metamaterials. Tianjin: Nankai University; 2013.
3. Wang M, Zhong K, Liu C, Degang X, Shi W, Yao J. Optical coefficients extraction from terahertz time-domain transmission spectra based on multibeam interference principle. Opt Eng. 2017;56(4): 044101.
4. Naftaly M, Miles RE. THz-TDS for material characterization. Proc IEEE. 2007;95(8):1658–65.
5. Dai J, Zhang J, Zhang W, et al. THz-TDS characterization of the far-infrared absorption and index of refraction of high-resistivity, float-zone silicon. J Opt Soc Am B. 2004;21(7):1379–86.
6. Kuzel P, Nemec H, Kadlec F, et al. Gouy shift correction for highly accurate refractive index retrieval in time-domain terahertz spectroscopy. Opt Express. 2010;18(15):15338–48.
7. Sommer S, Raidt T, Fischer BM, et al. THz-spectroscopy on high density polyethylene with different crystallinity. J Infrared Millim Terahertz Waves. 2015;37(2):189–97.
8. Ioachim P, Rafal W, Martin K. Highly accurate optical material parameter determination with THz time-domain spectroscopy. Opt Express. 2007;15(7):4335–50.

Chapter 4
Application of THz-TDS in the Characterization of Crystalline Materials

Optical parameters of crystals, including reflectivity, transmittance, refractive index, extinction coefficient, absorption coefficient, and dielectric constant, are characterization parameters of the macroscopic physical properties of crystals [1]. They are influenced by the distribution of molecules and atoms within the crystal, lattice vibration distribution, and electronic band transitions. Therefore, optical parameters serve as a link between the macroscopic characteristics and microscopic states of a crystal. By measuring the optical parameters of crystalline materials at the macroscopic level, we can indirectly infer the microscopic characteristics and mechanisms of the material. There are many connections between the optical parameters of crystals; they are not independent. For example, the famous Fresnel equation connects the refractive index with transmittance and reflectivity. Knowing the refractive index allows us to calculate the other two parameters. This provides options for practical experimental measurements. During experiments, we can choose the most convenient optical parameter to measure based on the actual situation and experimental conditions. Then, using the relationships with other optical parameters, we can calculate the required constants.

The optical parameters of crystals have dispersion characteristics and are closely related to the frequency of incident light. Therefore, the same material may exhibit completely different optical characteristics for incident light of different frequency bands. Hence, for a particular crystalline material, it is necessary to measure its optical parameters in different frequency bands. We cannot directly use the optical parameters from other frequency bands. Generally, the optical characteristics of crystalline materials in the microwave and light wave frequency bands differ significantly. For the transition frequency band in the terahertz range, the optical parameters of the material may exhibit characteristics different from those of the microwave and optical frequency bands on either side, possessing their own unique properties.

The optical parameters of different crystalline materials in the terahertz frequency band have significant practical application value. On one hand, many crystalline materials, such as gallium selenide (GaSe), zinc germanium phosphide (ZGP),

T. Chen et al., *Terahertz Time Domain Spectroscopy*,
https://doi.org/10.1007/978-981-95-2605-5_4

lithium niobate (LN), and potassium titanyl phosphate (KTP) [2], are often used in the generation and detection of terahertz waves. Their optical parameters, especially the absorption coefficient and refractive index, provide guidance for the selection of application bands, phase matching calculations, and crystal cutting design. On the other hand, for terahertz-transparent crystals like quartz, which are often used as terahertz output windows and terahertz lenses, parameters such as refractive index and absorption coefficient are crucial for their design as optical components. Studying their parameters in relation to temperature, stress, etc., can lay the foundation for the application design of terahertz components in different scenarios.

THz-TDS is a method used to measure the dynamic processes of materials, obtaining various characteristics of materials quickly with femtosecond to picosecond short time intervals [3, 4]. THz-TDS can be applied in material science to study the complex refractive index, dielectric constant, birefringence, electro-optic effect, and magneto-optic effect of materials in the terahertz frequency band. In a THz-TDS system, a broadband terahertz pulse is typically generated by a femtosecond laser pulse in a photoconductive antenna. After passing through the sample, this broadband terahertz pulse is detected in amplitude and phase by a delayed detection pulse, thereby directly obtaining the complex dielectric constant of the sample. In addition to terahertz emitters and detectors, other terahertz devices such as polarization beam splitters, converters, and filters are also very important for terahertz technology. High-speed THz-TDS systems provide a very effective way to study materials by measuring their absorption, transmission, and reflection, and further obtaining the refractive index, absorption coefficient, dielectric parameters, and other optical parameters of the materials. THz-TDS can be used to detect the spectral characteristics of crystals in the terahertz band [5].

The optical characteristics of crystalline materials in the terahertz band form the basis for their application in terahertz radiation sources, terahertz polarization devices, terahertz windows, and other fields. Compared to far-infrared Fourier spectrometers, ellipsometers, and other testing equipment, THz-TDS systems have advantages such as coherent detection, high signal-to-noise ratio, fast sampling speed, wide coverage, and low sample requirements. They can conveniently obtain the transmittance, reflectance, refractive index, absorption coefficient, dielectric parameters, etc., of crystalline materials, providing an efficient and convenient means of characterizing optical parameters in the terahertz band. In this chapter, we will focus on α/β-BTM, $AgGaS_2$, $ZnGeP_2$. The application of THz-TDS in crystal property testing and analysis is demonstrated using three types of crystals and transmission measurement methods. This includes different lattice structures of the same type of crystal, different temperature conditions of the same crystal, and different growth methods of the same crystal, highlighting the accuracy and practicality of time-domain spectroscopy in crystal parameter measurement. However, when time-domain spectroscopy is used for reflective measurement of material parameters, due to the reliance on a gold mirror to provide a reference reflection signal, replacing the metal reflection mirror with the sample to be tested will introduce a phase difference, leading to larger errors in the measurement of refractive index and absorption coefficient in reflective measurements, limiting the application of time-domain spectroscopy systems in characterizing non-transparent material parameters.

4.1 Characterization of A/β-BTM Crystal Terahertz Band Properties [6]

$BaTeMo_2O_9$ (BTM) is a mid-infrared nonlinear optical crystal material first grown by Shandong University, with excellent second and third order nonlinear optical properties. BTM crystals can be grown using a solvent-assisted method. When the concentration is low, the system can crystallize a single phase of α-BTM crystal, and when the concentration is high, it can crystallize a pure phase of β-BTM crystal. The α-BTM has an orthorhombic structure, and the β-BTM has a monoclinic structure. Both types of crystals are excellent Raman media, with high Raman gain and appropriate Raman shift, achieving good results in the generation of lasers in the 1178–1531 nm wavelength range [7, 8]. On the other hand, the β-BTM crystal has a larger second-order nonlinearity ($d_{31} = -9.88$ pm/V) than α-BTM and has successfully achieved high-efficiency yellow laser output through self-doubling Raman conversion. In particular, β-BTM has good physical properties, a large nonlinear coefficient (for the doubling process of 1064–532 nm, the d_{eff} value of β-BTM is 3.17 times that of $LiNbO_3$), high Raman gain, a wide transmission range (0.4–5.5 μm), and a high damage threshold (>500 MW/cm^2).

Both α- and β-BTM crystals are grown by the top-seeded melt method. The dielectric axes x, y, and z of the orthorhombic α-BTM crystal coincide with the crystallographic axes b, c, and a, respectively, but in the monoclinic β-BTM crystal, only one dielectric axis (z) overlaps with the crystallographic axis (b), the other two orthogonal dielectric axes x and y are located in the ac plane. The angle between the y-axis and the a-axis is 25.5°, and the angle between the x-axis and the c-axis is 24.603°. To characterize its optical properties within the terahertz band, thin slices (about 1 mm for α-BTM, about 0.6 mm for β-BTM) were cut from the bulk single crystal and polished. Since the crystals under study are biaxial crystals, sample transmission directions along the three principal dielectric axes (x, y, and z) were designed, as shown in Table 4.1.

Table 4.1 The α-BTM and β-BTM crystal settings

α-BTM		β-BTM	
Direction	Thickness (mm)	Direction	Thickness (mm)
x-cut	1.064	*x*-cut	0.609
y-cut	1.032	*y*-cut	0.630
z-cut	1.026	*z*-cut	0.642

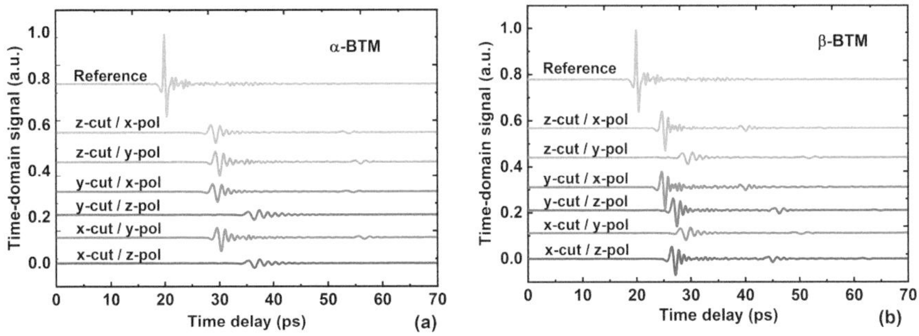

Fig. 4.1 Terahertz time-domain signal of α-BTM and β-BTM crystals

The time-domain signals of the measured samples and their corresponding reference signals are shown in Fig. 4.1. The smaller peaks are caused by internal reflections between the two parallel surfaces of the sample. Each sample measures terahertz polarization along two axes, and there are six time-domain signal curves in total for α and β-BTM crystals. Even if the crystal orientation is different (ignoring thickness differences), when the terahertz polarization is along a certain axis, it has the same signal, which means that the dielectric response of the crystal depends on the field polarization rather than the propagation direction. The total time delay of the β-BTM samples is smaller because they are much thinner. It should also be noted that the time delays of the α-BTM samples in x and y polarizations are similar, both of which are much smaller than the time delay in z polarization, indicating that the refractive index follows $n_{\alpha x} \approx n_{\alpha y} < n_{\alpha z}$. However, for β-BTM samples, the time delays of x, y, and z polarizations are significantly different, and y polarization has the maximum delay, indicating $n_{\beta x} < n_{\beta z} < n_{\beta y}$.

The terahertz time-domain waveforms are transformed by Fast Fourier Transform to obtain their spectral distribution as shown in Fig. 4.2. The differences in transmittance along different axes come from the anisotropy of the refractive index and absorption rate. In the low-frequency range, the transmittance is mainly determined by surface reflection related to the refractive index, while the transmittance in the high-frequency part depends on the absorption of phonon modes. The two obvious dips in the y-polarization transmission curve (located at 1.04 THz and 1.22 THz) correspond to the two strong absorption phonon modes in this range. According to the transmission spectrum, the reliable measurement range of α-BTM in this system is 0.2–1.3THz (x-axis), 0.2–1.5THz (y-axis), and 0.2–1.0THz (z-axis); the reliable measurement range of β-BTM in this system is 0.2–1.8THz (x-axis), 0.2–1.2THz (y-axis), and 0.2–1.45THz (z-axis).

The refractive index results are given in Fig. 4.3, where the solid line represents the experimental results and the dashed line represents the calculated results of the Sellmeier equation fitting. Consistent with the prediction of the time-domain signal given in Fig. 4.2a, the difference between x and y polarizations is small (0.6 THz when $n_{\alpha y}$ -$n_{\alpha x}$ = 0.13). For α-BTM, the birefringence between y and z polarizations is large (at 0.6 THz, $n_{\alpha z}$ -$n_{\alpha y}$ = 1.75). Conversely, for β-BTM, all three polarizations

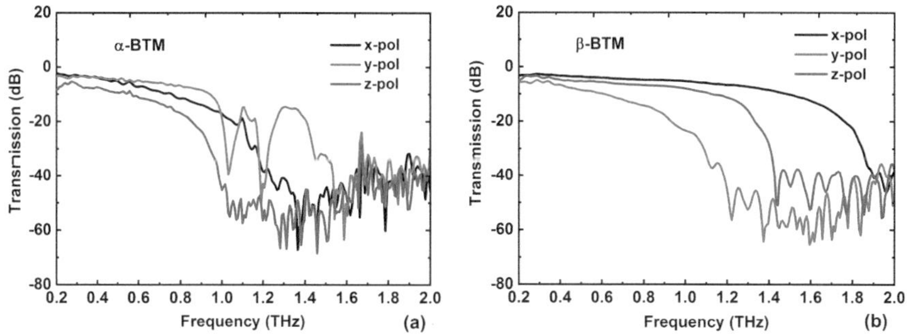

Fig. 4.2 Frequency domain transmission spectrum of α-BTM and β-BTM crystals

($n_{\beta y} - n_{\beta x} = 1.93$ and $n_{\beta z} - n_{\beta y} = -0.97$ at 0.6 THz) have significant birefringence. Figure 4.4 also shows the difference between n_x and n_y of the two crystals, where two z-cut samples were used, and the terahertz polarization has a certain angle with the main axes x or y. The signal delay of the time-domain signal transmitted through the β-BTM sample (~4.1 ps) is very consistent with the phase difference calculated by birefringence ($n_y - n_x$) and sample thickness, while for α-BTM with much smaller birefringence, the signal splitting can be ignored.

Since the existing BTM crystal refractive index equations from the visible to the mid-infrared range are not applicable in the terahertz range, to provide a reference for the application of BTM crystals in the terahertz range, the refractive indices of α- and β-BTM crystals in the measurement band are fitted with the Sellmeier equation to obtain their refractive index curve equations in the corresponding frequency band:

$$n_i^2 = A + \frac{B\lambda^2}{\lambda^2 - C} + \frac{D\lambda^2}{\lambda^2 - E} \tag{4.1}$$

where i represents the three main axes x, y, and z, λ is the wavelength, in μm, and A, B, C, D, and E are the Sellmeier parameters given in Table 4.2. The comparison between the fitting curve and the measurement results is shown in Fig. 4.3. The two

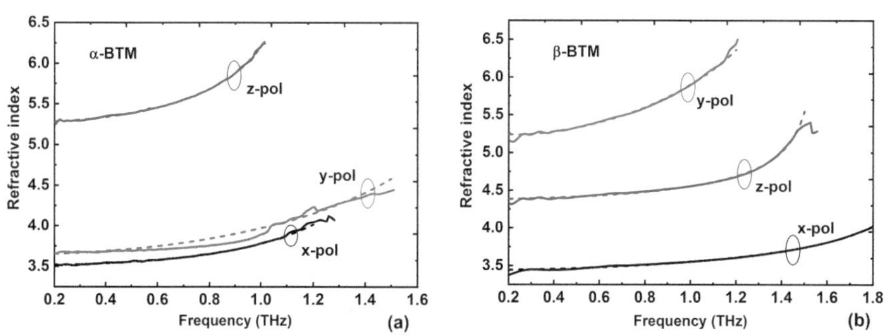

Fig. 4.3 Refractive indices of α-BTM and β-BTM crystals

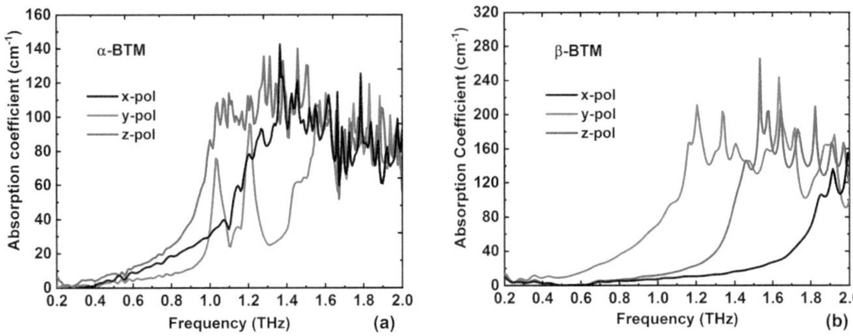

Fig. 4.4 Absorption coefficients of α-BTM and β-BTM crystals

fluctuations in the y-polarization refractive index curve of α-BTM in the figure are caused by the phonon modes at 1.04 and 1.22 THz, causing a jump in the dielectric curve.

The absorption coefficient is derived from the complex amplitude and refractive index of the terahertz field, as shown in Fig. 4.4. Due to the high frequency approaching the far-wing absorption of the phonon mode, the absorption coefficient increases with frequency. On the surface, the relationship between the absorption coefficients of different dielectric axes appears random, for α-BTM it is usually $\alpha_{\alpha z} > \alpha_{\alpha x} > \alpha_{\alpha y}$, for β-BTM it is usually $\alpha_{\beta y} > \alpha_{\beta z} > \alpha_{\beta x}$, but both types of crystals show good consistency from the crystallographic axes ($\alpha_a > \alpha_b > \alpha_c$). The results show that terahertz polarization along the c-axis has the widest transmission range, as both crystals are c-oriented two-dimensional layered structures, where the coupling between the electric field and chemical bonds is weakest.

The crystal molecular vibration modes can be analyzed through the absorption spectrum. On one hand, α-BTM has an orthogonal crystal structure $Pca_{21}(C_{2v}^5)$, according to group theory calculations, it should be in the $78A_1 + 78B_1 + 78A_2 + 78B_2$ modes, where A_1, B_1, and B_2 modes have both Raman and infrared activity.

Table 4.2 Sellmeier coefficients of α-BTM and β-BTM crystals

Sellmeier coefficients	α-BTM			β-BTM		
	n_x	n_y	n_z	n_x	n_y	n_z
A	3.6	4.7	6.207	1.6	5	4
B	4.509	−2	14.4	9.291	1.387	15.35
C	303.5	−49.69	24.13	45.43	350.7	-6.556×10^5
D	4.217	10.57	7.078	0.9495	12.79	22.72
E	−77.97	166.1	536.5	202.1	46.54	235.8
Validity range	0.2–1.3 THz	0.2–1.5 THz	0.2–1.0 THz	0.2–1.8 THz	0.2–1.2 THz	0.2–1.45 THz

Due to the bandwidth and signal-to-noise ratio limitations of the THz-TDS system, most modes are not reflected in Fig. 4.4a. However, under the y-polarized terahertz field, two A1 modes at 1.04 THz and 1.22 THz can be observed, indicating that the dipoles generated by the vibrations are along the y-axis (c-axis). These two modes were observed for the first time. On the other hand, β-BTM has a monoclinic P_{21} (C_2) structure, and according to group theory calculations, it should be in the 39A + 39B modes, where both A and B modes have Raman and infrared activity. When the terahertz field is along the z-axis, the corresponding modes are excited, corresponding to strong absorption peaks. However, since the vibrations of the B mode are in the x–y plane but not parallel to any axis, even if the absorption peaks of x and y polarization are recorded, these phonon states cannot be directly reflected in Fig. 4.4b. Based on the above measurements, the optical properties of α- and β-BTM crystals in the 0.2–2 THz frequency range were studied using THz-TDS. The refractive indices of the two crystals were measured, and the Sellmeier equation within their transparent frequency range was obtained by fitting the measurement data. It was found that the β-BTM crystal has a larger birefringence in the x–y plane than the α-BTM crystal. The characterization accurately reflects the structural differences between the orthorhombic and monoclinic systems. By analyzing the absorption coefficient, two lattice vibration modes at 1.04 THz and 1.22 THz along the y-axis of α-BTM were found. Moreover, corresponding to their c-oriented two-dimensional layered structure, both crystals have lower absorption along the c-axis. The dielectric properties of the crystal obtained through THz-TDS provides reference parameters for the application of materials in the terahertz frequency band, providing a theoretical basis for predicting the application of crystals in the terahertz frequency band. Meanwhile, the dielectric parameter differences of the two different crystal forms can be compared, from which we can analyze the changes in the crystal structure and discuss the properties of the vibration energy levels and bands corresponding to their crystal structures.

4.1.1 Characterization of $AgGaS_2$ Crystal Terahertz Band Properties at Different Temperatures [9]

Silver gallium sulfide ($AgGaS_2$) is an important infrared nonlinear optical material and a typical I-III-VI compound with a chalcopyrite structure, belonging to the 42 m point group [10–12]. $AgGaS_2$ is a birefringent crystal with a transparency range of 0.49–12.5 μm, a high nonlinear coefficient ($d_{36} = 13 \pm 2$ pm/V), and a high damage threshold (>50 MW/cm^2). Various nonlinear optical devices based on difference frequency generation (DFG), optical parametric generation (OPG), and optical parametric oscillation (OPO) [13–15] have been realized based on $AgGaS_2$. Kato achieved efficient generation of continuous tunable high-power mid-infrared radiation in the 5–11 μm range based on DFG. Elsässer et al. achieved a wide-tunable OPG of 1.2–10 μm with a conversion efficiency of up to 10%. Vodopyanov et al. built a tunable $AgGaS_2$ OPO with a tuning range of 3.9–11.3 μm, a maximum slope efficiency of 41%, and a corresponding quantum conversion efficiency of 22%. Broadband terahertz generation [16–18] was also achieved through DFG and optical rectification (OR) using the wide transparency range, high nonlinearity, and good physical properties of $ZnGeP_2$, $CdSiP_2$, and $AgGeSe_2$ crystals. As one of the most representative chalcopyrite crystals, $AgGaS_2$, which is the best choice for 1 μm laser pumping, has extremely high potential in the terahertz band. Therefore, systematic characterization and analysis of the optical properties of this crystal in the terahertz range are of great reference significance for evaluating phase matching conditions, output performance, etc. To characterize the birefringence of $AgGaS_2$, a 10 mm × 10 mm × 1.6 mm (100) single crystal slice was used as the sample. The changes in the dielectric properties of the crystal at different temperatures are discussed. The sample is placed in a specially designed temperature-controlled sample holder, controlling the temperature from room temperature to 230 °C. By rotating the sample 90° around the light direction, the dielectric properties of the crystal under different polarization directions are measured, thus discussing the changes in the phonon modes of the AGS crystal at different polarizations.

Measurements were taken at 30 °C, 90 °C, 120 °C, 160 °C, 190 °C, and 230 °C. Figure 4.5 shows the time-domain signal and corresponding Fourier transform spectrum of $AgGaS_2$. To avoid the influence of temperature changes on the air background, reference signals at each temperature were recorded, and it was found that the water absorption peak near 1.7 THz decreases with increasing temperature. The time delays (peak values of signal pulses) of the two polarization directions are similar, indicating that the main refractive indices of ordinary (o-wave) and extraordinary (e-wave) waves in the terahertz range are close. From the time-domain signal, as the temperature increases, the pulse signal delay continuously increases. When the temperature is 30 °C, $\Delta t = 0.01$ ps, and when the temperature is 230 °C, $\Delta t = -0.01$ ps, the delay difference has changed sign, mainly due to the flip between birefringence. Figure 4.2 presents the comparison results of the spectral density of the $AgGaS_2$ sample and the sample without $AgGaS_2$. The transmission differences of different polarization states originate from the anisotropy of crystal refractive index and absorption. At the low-frequency end, transmission is mainly determined by surface reflection related to the refractive index, while at the high-frequency part, it mainly depends on the absorption wing of optical phonons. The significant tilt angles at 1.03 THz ($E \perp c$) and 1.87 THz ($E \parallel c$) on the curve are related to strong absorption caused by two phonon modes. The decrease in transmittance during heating is also significant, indicating that lattice vibrations cause significant changes in refractive index and absorption.

Figure 4.6 shows the temperature-dependent characteristics of the transmittance of $AgGaS_2$ under different polarizations. The transmission differences of different polarization directions originate from the anisotropy of crystal refractive index and absorption. At the low-frequency end, transmission is mainly determined by surface reflection related to the refractive index, while at the high-frequency part, it mainly depends on the absorption wing of optical phonons. The significant tilt angles at 1.03 THz ($E \perp c$) and 1.87THz ($E \parallel c$) on the curve are related to strong absorption

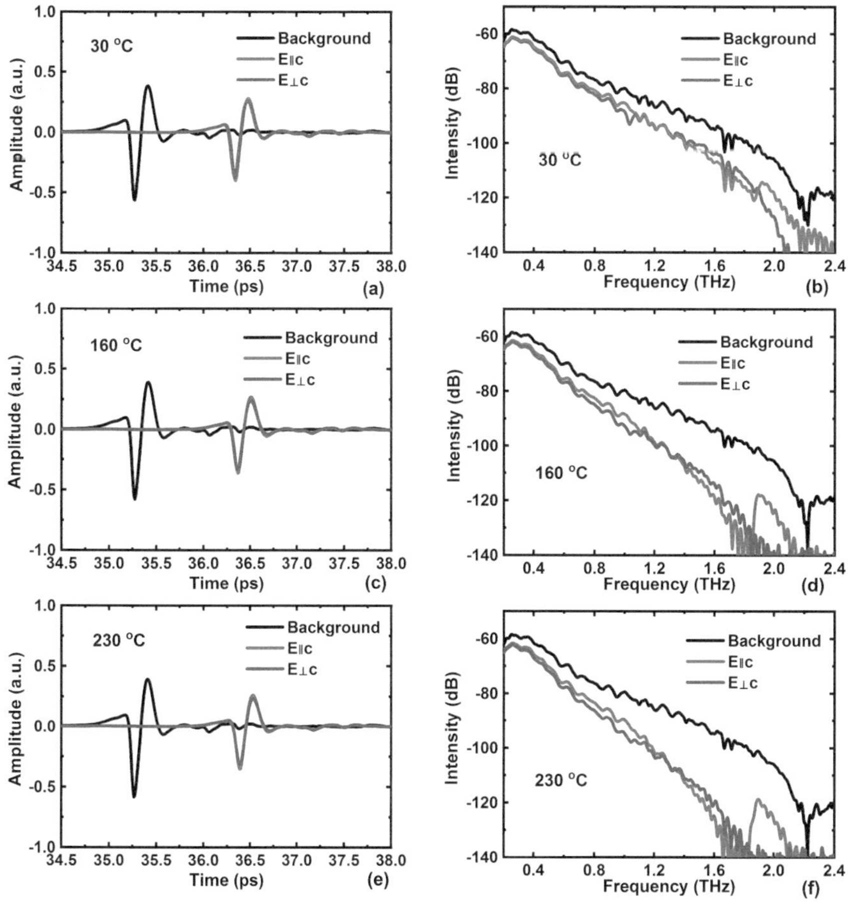

Fig. 4.5 Terahertz time-domain signal and corresponding fourier transform spectrum of $AgGaS_2$ at different temperatures

caused by two phonon modes. The decrease in transmittance during heating is also significant, indicating that lattice vibrations cause significant changes in refractive index and absorption.

The refractive index measurement results are shown in Fig. 4.7a, b. Consistent with the prediction of the time-domain signal, the difference between n_o and n_e is not significant within the effective frequency range. Consistent with the predictions from the time domain signals, there is little difference between n_o and n_e within the valid frequency range. At a certain temperature, n_o is larger than n_e in the low-frequency range to be a negative uniaxial crystal, and it inverts at around 1.5 THz where the optical phonon introduces a dramatic change. Basically, the refractive index increases linearly with increasing temperature, and the change in n_o is greater than that in n_e. The change in refractive index with temperature is determined by

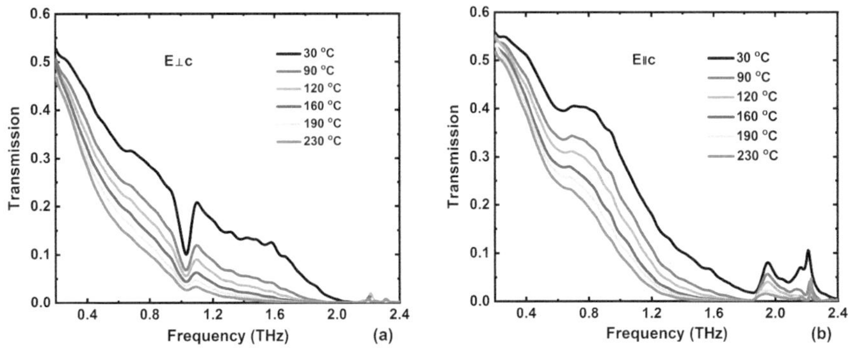

Fig. 4.6 Temperature-dependent transmission of $AgGaS_2$ under different polarizations: **a** E⊥c; **b** E||C

two factors: the thermal expansion of the crystal and the dielectric response induced by phonon vibrations. $AgGaS_2$ has a smaller thermal expansion coefficient in the plane perpendicular to the c-axis, and the strong vibration of optical phonons (for example, the phonon mode at 94 cm^{-1}) plays a decisive role in the high-frequency region. Figure 4.7c shows that, away from the frequency of optical phonons, as the temperature increases, the birefringence of $AgGaS_2$ mainly decreases. This may be mainly caused by the crystal structure composed of Ag–S bonds and Ga-S bonds. The electronegativity difference of the Ga-S covalent bond is larger than that of the Ag–S bond, so the Ga-S bond has a shorter bond length, stronger bond energy, stronger covalency, and better thermal stability. When the temperature changes, the unbalanced changes of the two bonds cause the tetragonal lattice to distort, thereby causing changes in birefringence.

Based on the measurement data, the Sellmeier equation of the $AgGaS_2$ crystal in the measurement range can be obtained through data fitting:

$$n_i^2 = A_i + \frac{B_i\lambda^2}{\lambda^2 - C_i} \tag{4.2}$$

where i represents the two polarization states, λ is the wavelength, and A, B, and C are fitting parameters. To obtain the temperature dependence of the refractive index, the first step is to fit all the measured values at each temperature, from which it is found that the relationship between the fitting parameters and temperature is close to linear. Then, by adding a linear temperature coefficient to the fitting parameters, the temperature dependence of the refractive index is introduced into the equation. The established temperature-related Sellmeier equation is as follows:

$$\begin{aligned} n_o^2(\lambda, T) &= 8.48 + 0.00294 \times T \\ &\quad + \frac{0.2825 \times \lambda^2}{\lambda^2 - \left(1.38374 \times 10^4 + 0.00204 \times 10^4 \times T\right)} \end{aligned} \tag{4.3a}$$

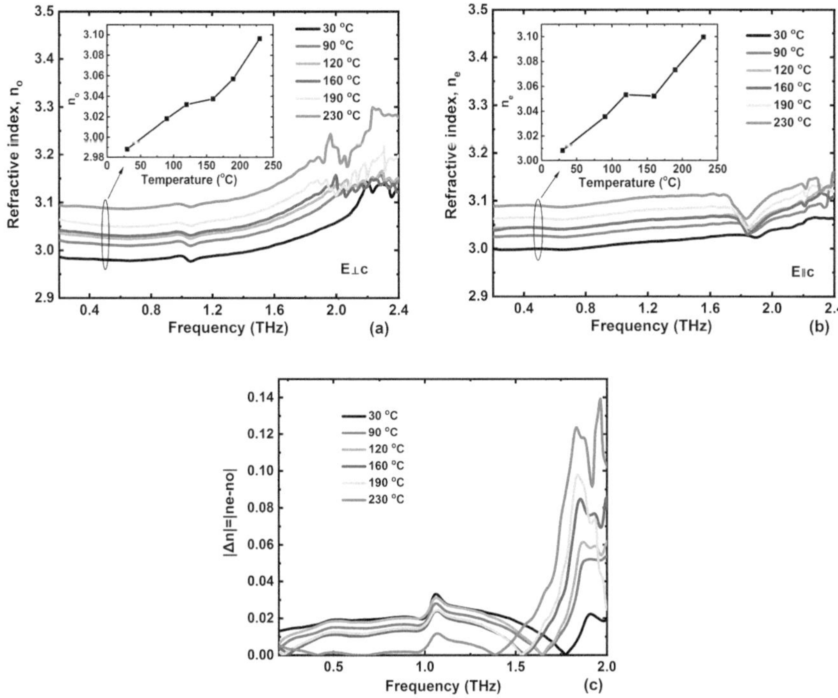

Fig. 4.7 Temperature-dependent refractive index characteristics of $AgGaS_2$: **a** E⊥c (n_o); **b** E∥C (n_e); **c** Birefringence $|\Delta n| = |n_e - n_o|$. The insets in **a** and **b** show the change in refractive index with temperature at 0.5 THz

$$n_e^2(\lambda, T) = 7.55542 + 5.8226 \times 10^{-4} \times T + \frac{(0.00207 \times T + 1.34739) \times \lambda^2}{\lambda^2 - (3712.8 - 6.7644 \times T)} \tag{4.3b}$$

The comparison between the fitting equation and the measured data is shown in Fig. 4.8, where the solid line and the scatter points represent the calculated values and the measured values of the Sellmeier equation, respectively. Except for the strong absorption at the phonon mode, the overall refractive index deviation is within 0.01. The refractive index related to wavelength and temperature within this wavelength range is a reliable temperature-variable Sellmeier equation, which is very important for calculating the phase-matching conditions of nonlinear frequency conversion of the crystal in this frequency range.

The absorption coefficient of the terahertz band of the $AgGaS_2$ crystal is shown in Fig. 4.9. The insect shows the change in the absorption coefficient at 1 THz with temperature. Similar to the refractive index, the absorption anisotropy in both polarization directions is weak, especially in the low-frequency range far from the phonon mode. At a fixed temperature, the absorption coefficient usually increases

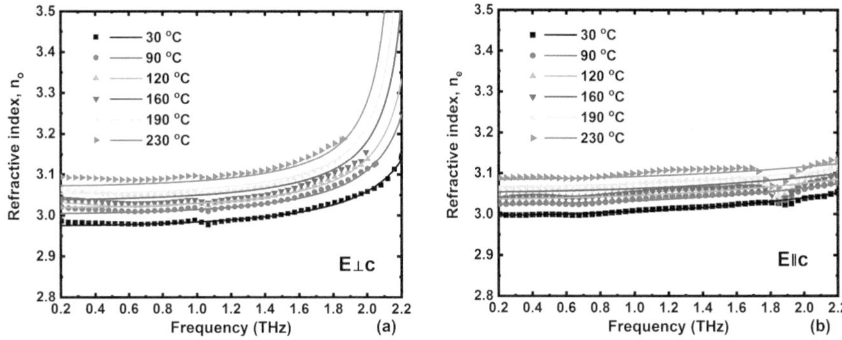

Fig. 4.8 Refractive index of $AgGaS_2$ calculated from the fitted temperature-dependent Sellmeier equation: **a** comparison of measurements for E⊥c (n_o); **b** E∥c (n_e)

with frequency because the strong phonon vibration in the far wing exceeds the measurable range. Considering that the $AgGaS_2$ crystal has a point group $\bar{4}2$ m (D2d) structure, group theory predicts $A_1 + 2A_2 + 3B + 4B_2 + 7E$ Brillouin zone center modes [19], where the phonon vibration modes are Γacoustic $= B_2 + E$ and the optical phonon vibration modes are $\Gamma_{optical} = A_1 + 2A_2 + 3B_1 + 3B_2 + 6E$. Previous calculations showed that the infrared active modes are $3B_2 + 6E$ [19]. Measurements show that the B_2 mode with the dipole generated by vibration along the *c* axis has a center frequency of 1.86 THz (62 cm^{-1}), and the E mode vertical to the *c* axis has a center frequency of at 1.03 THz (34 cm^{-1}), both at room temperature. As the temperature increases, the lattice vibration intensifies, reflected in the increase in the absorption coefficient and the broadening of the absorption peak.

Another effect of temperature increase on phonons is the shift in the center frequency of the absorption peak, indicating that the $AgGaS_2$ crystal undergoes phonon softening as the temperature rises. This is due to the thermal expansion increasing the atomic spacing, resulting in a smaller force constant, which in turn lowers the corresponding center frequency [20]. Figure 4.10 shows the detailed

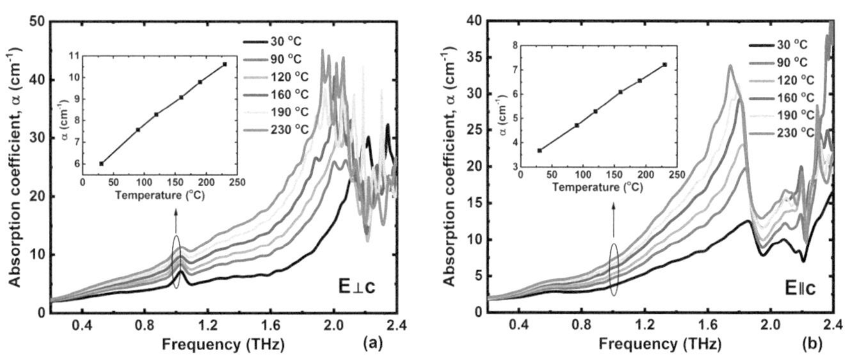

Fig. 4.9 Temperature-dependent absorption coefficient spectra of $AgGaS_2$ crystals **a** E⊥c **b** E∥c

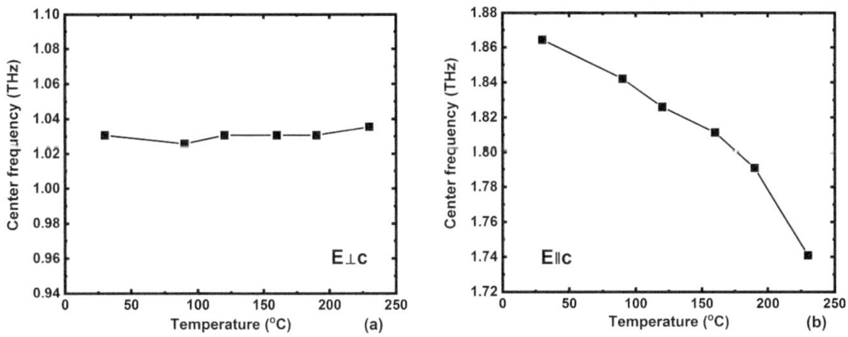

Fig. 4.10 Relationship between central phonon vibration frequency and temperature: **a** E⊥C **b** E||C

changes in the modes at 1.03 THz and 1.86 THz with temperature. Phonon softening is particularly noticeable only when the magnetic field is along the c-axis (B_2 mode, E||c). Its center frequency moves from 1.86 THz at 30 °C to 1.74 THz at 230 °C, and then almost linearly decreases. The center frequency of the other mode (E mode, E⊥c) does not change significantly. The difference in phonon softening in $AgGaS_2$ related to thermal expansion anisotropy also provides evidence for the previous discussion about the difference in refractive index changes corresponding to the two polarizations.

To further study phonon softening, we also used the iHR320 Raman spectrometer manufactured by HORIBA Scientific (wavelength range: 150 nm–40 μm, resolution: 0.06 nm) to measure the unpolarized Raman spectrum of the $AgGaS_2$ crystal at temperatures of −17 °C, 30 °C, and 70 °C, as shown in Fig. 4.11. The vibration modes at 34 cm^{-1}, 94 cm^{-1}, 160 cm^{-1}, 229 cm^{-1}, and 389 cm^{-1} are E modes, and the strongest vibration mode at 290 cm^{-1} is an A_1 mode. Five E modes with distinct peaks were selected from the Raman spectrum and their half-widths were obtained by Gaussian fitting. Due to stronger phonon vibration, the intensity of the Raman spectrum tends to increase with temperature. The blue shift and broadening of the Raman peaks confirm the phonon softening of the E mode during heating, although this change is not as obvious as the B_2 mode. At 34 cm^{-1}, the phonon softening of the mode can also be distinguished by the Raman spectrum.

The optical properties of a crystal are the result of the interaction of phonon modes, each of which can be considered as a Lorentz oscillator. Typically, dielectric properties can be represented by a multi-oscillator model [21, 22]:

$$\varepsilon(\nu)=(n+i\kappa)^2=\varepsilon_\infty+\sum_j\frac{S_j}{\nu_j^2-\nu^2-i\Gamma_j\nu} \tag{4.4}$$

In the formula, ε(ν) is the complex dielectric constant, ν is the wave number, ν_j is the eigenfrequency of the corresponding phonon mode in $AgGaS_2$, ε_∞ is the high-frequency dielectric constant, S_j is the oscillator strength, and Γ_j is the damping coefficient. The parameters in the theoretical model at room temperature are shown

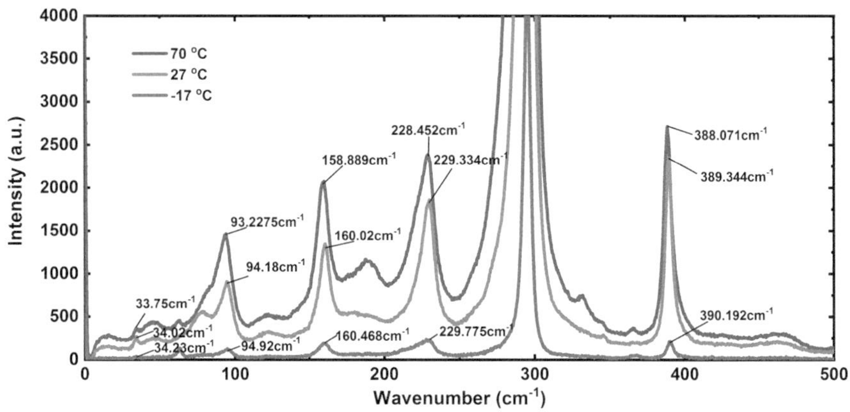

Fig. 4.11 Raman spectra of $AgGaS_2$ at different temperatures

in Table 4.3 [23, 24]. The phonon mode data at 34 cm^{-1}, 62 cm^{-1}, and 95 cm^{-1} were obtained by fitting the measured absorption spectra, while other data are from simulation results [24].

When considering only the influence of the E mode and B_2 mode on dielectric properties, the refractive index and absorption coefficient calculated using the multi-oscillator model are shown in Fig. 4.12. The calculation results show that when E⊥c, the theoretical values of dielectric properties align well with the measured values, whereas when E||c, there is a certain discrepancy. This indicates that when E⊥c, the dielectric properties of the crystal are mainly influenced by the E mode, thus the calculated values using the multi-oscillator model are more accurate. However, when E||c, the dielectric properties of the crystal are influenced not only by the B_2 mode but possibly by multiple modes. Overall, the Lorentz model can be applied to describe the dielectric characteristics of $AgGaS_2$ crystals.

Based on the above measurement data, the optical properties (refractive index and absorption coefficient) of AGS at different temperatures in the 0.2–2 THz frequency range were studied through TDS. The refractive index curve equations and dielectric

Table 4.3 Fitting parameters of the Lorentz model for E modes and B_2 modes

E modes (ε_∞ = 5.9 at 1.2 μm)			B_2 modes (ε_∞ = 5.7 at 1.2 μm)		
ν_j (cm^{-1})	S_j (cm^{-1})	Γ_j (cm^{-1})	ν_j (cm^{-1})	S_j (cm^{-1})	Γ_j (cm^{-1})
34	0.00054×10^4	4	61	0.0312	9
95	0.852×10^4	11.5	213.6	9.53×10^4	5.66
160	3.79×10^4	48.5	364.8	9.86×10^4	13.8
227	3.57×10^4	13.49			
343	1.55×10^5	7.49			
387	3.85×10^4	3.92			

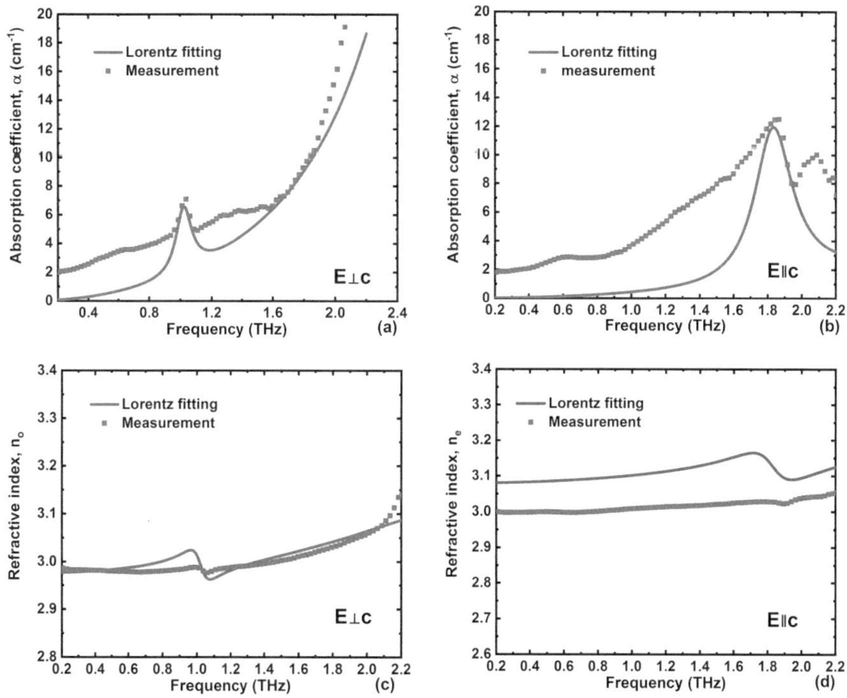

Fig. 4.12 Comparison of Lorentz fitting model with measured absorption coefficient and refractive index at room temperature. **a** and **c**: E⊥C; **b** and **d**: E||C

models of at different temperatures were obtained, laying the basis for applications of AGS in the terahertz band.

4.1.2 Terahertz Spectroscopy Measurement of Zinc Germanium Phosphide (ZGP) Crystals [25]

Zinc germanium phosphide ($ZnGeP_2$, abbreviated as ZGP) is a tetragonal crystal with a chalcopyrite point group structure [26]. ZGP crystals have suitable birefringence characteristics for phase matching and exhibit excellent optical, mechanical, and thermal properties, along with a relatively high laser damage threshold. Compared to most infrared crystals such as gallium selenide (GaSe), gallium phosphide (GaP), and gallium arsenide (GaAs), ZGP crystals possess much higher nonlinear optical quality [27]. Additionally, due to their broad transmission band of 0.74–12 μm and low absorption coefficient in the terahertz band, ZGP crystals are widely used in terahertz generation and detection. Experiments on generating continuously tunable single-frequency terahertz waves through difference frequency generation (DFG) using ZGP crystals have been reported [28, 29], with a maximum average output power reaching 2 mW [30]. In the field of generating broadband terahertz pulses pumped by near-infrared femtosecond laser based on collinear phase-matched optical rectification, ZGP crystals show significant advantages over other nonlinear materials [31, 32], such as lower nonlinear absorption, larger second-order nonlinear coefficients, smaller velocity mismatch in the near-infrared band, and smaller free carrier absorption below 3 THz. Furthermore, the electro-optic detection capability of ZGP crystals in the terahertz band indicates that ZGP has a larger phase modulation bandwidth compared to commonly used zinc telluride (ZnTe), zinc selenide (ZnSe), and gallium phosphide (GaP) crystals [18].

The application of ZGP crystals in the terahertz band requires a detailed and comprehensive understanding of their properties in this frequency range. However, compared to studies on this crystal in the near-infrared and far-infrared bands [33, 34], the optical properties of ZGP crystals in the terahertz band have not been thoroughly investigated. Apart from a few studies on Raman and infrared spectroscopy, the spectral characteristics of ZGP crystals in the terahertz band have been rarely reported.

ZGP crystals can be obtained through horizontal gradient freezing (HGZ) and vertical gradient freezing (VGZ) techniques. Some characteristics of crystals grown by these two methods in the near-infrared and mid-infrared bands have also been reported [35]. It is known that ZGP crystals obtained by horizontal growth have an absorption coefficient of less than 0.01 cm^{-1} in the 3–8 μm band and still less than 0.03 cm^{-1} at 2 μm. For vertically grown crystals, these values are 0.02 cm^{-1} and 0.04 cm^{-1}, respectively, indicating that horizontally grown ZGP crystals are of relatively better quality, and they perform better in the conversion from 2 μm to 3–5 μm range, with the optical conversion efficiency reaching 56.2%. To characterize the optical properties of ZGP crystals grown by different methods in the terahertz frequency range, ZGP crystal samples of (100) and (001) orientations were prepared, with the aperture of 10 mm × 10 mm and the thickness ranging from 1.24 to 1.30 mm. The polarization direction of the terahertz beam was aligned with the principal axis of the crystal, with the o-polarized beam along the x-axis or y-axis, and the e-polarized beam along the z-axis.

Based on THz-TDS, the time-domain spectral signals of ZGP crystal (100) samples were acquired. After performing a Fourier transform on the time-domain signals, the corresponding frequency-domain signals were obtained, as shown in Fig. 4.13, which also includes the corresponding background reference signals. It is evident that the time-domain terahertz signals after two polarization directions of the crystals grown by both methods have pulse waveforms similar to the background waveforms, with reduced intensity. However, the signal time delay differs: firstly, for each sample, the e-polarized light has a longer time delay compared to the o-polarized light, suggesting that the overall refractive index of the e-polarized light is higher than that of the o-polarized light in this frequency range, consistent with the properties of a positive uniaxial crystal; secondly, the time delay difference between the o-polarized and e-polarized light of the vertically grown ZGP crystal is significantly larger than that of the horizontally grown crystal, indicating a higher birefringence in this frequency range. Considering the thicknesses of the horizontally and vertically grown ZGP samples are 1.240 mm and 1.290 mm, respectively, the overall birefringence for samples obtained by different growth methods is 0.019 and 0.033, respectively.

The refractive index and absorption coefficient of horizontally and vertically grown ZGP (100) samples under o and e polarizations are shown in Fig. 4.14. In this frequency range, there are no significant low-frequency phonons, so the refractive index exhibits a smooth curve. As inferred from the time-domain spectral signals above, the birefringence of ZGP crystals obtained by the vertical growth method is significantly larger than that obtained by the horizontal growth method. For example, at 1.5 THz, the birefringence of the vertically grown sample is 0.034, while that of the horizontally grown sample is 0.021. This phenomenon should be caused by different crystallization conditions, leading to different strains within the equivalent plane.

The measurement results indicate that the growth method has a significant impact on the crystal properties in the terahertz frequency range. Therefore, characterizing crystals grown by different methods using THz-TDS can effectively analyze the

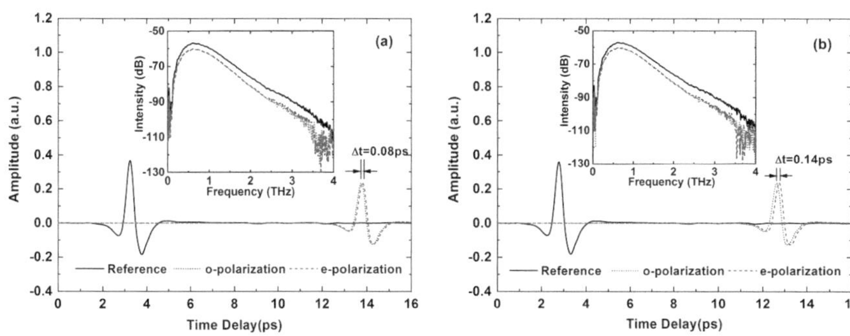

Fig. 4.13 Time-domain and Fourier-transformed frequency-domain spectrum of the (100) ZGP sample. **a** Horizontally grown; **b** Vertically grown

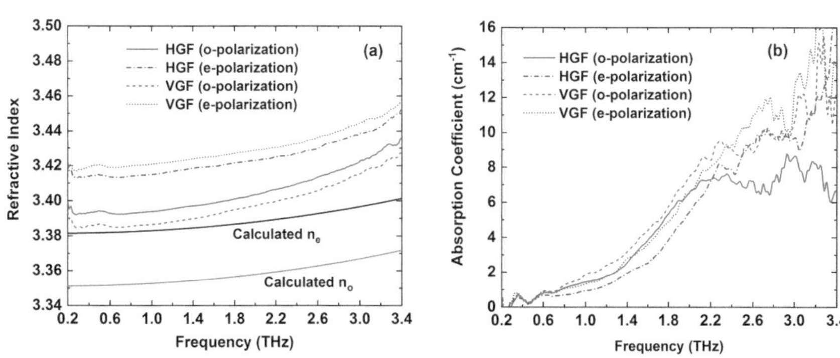

Fig. 4.14 Terahertz spectral measurements of HGF and VGF (100) ZGP crystals **a** refractive index; **b** absorption coefficient

influence of factors such as stress, growth temperature, quenching, and extraction processes on the properties of the resulting crystals, providing a reference for further optimizing growth procedure and improving crystal quality.

References

1. Rongchuan F. Solid state spectroscopy. University of Science and Technology of China Press;2003.
2. Luo Zhiwei G, Xin'an ZW, et al. Optical properties of GaSe:S crystals in terahertz frequency range. Opt Precis Eng. 2011;19(2):354.
3. Antsygin VD, Kaplun AB, Mamrashev AA, et al. Terahertz optical properties of potassium titanyl phosphate crystals. Opt Express. 2014;22(21):25436–43.
4. Liang Q, Wang S, Tao X, et al. Optical properties of $LiInSe_2$ in the THz frequency regime. Opt Mater Express. 2014;4(7):1336–44.
5. Duvillaret L, Garet F, Coutaz JL. A reliable method for extraction of material parameters in THz-TDS. IEEE J Sel Top Quantum Electron. 1996;2(3):739–46.
6. Qiao H, Zhong K, Yan C, Liu Y, Tang L, Liu K, Gao Z, Tao X, Li J, Degang X, Wang Y, Shi W, Yao J. $BaTeMo_2O_9$ crystals: optical properties and applications in the terahertz range. Opt Mater Express. 2019;9(11):4390–8.
7. Bai F, Wang QP, Tao XT, Li P, Zhang XY, Liu ZJ, Shen HB, Lan WX, Gao L, Gao ZL, Zhang JJ, Fang JX. Eye-safe Raman laser based on $BaTeMo_2O_9$ crystal. Appl Phys B. 2014;116(2):501–5.
8. Liu SD, Zhang JJ, Gao ZL, Wei L, Zhang SJ, He JL, Tao XT. Generation of 1.3 μm and 1.5 μm high-energy Raman radiations in α-BTM. Opt Mater. 2014;36(4):760–763.
9. Qiao H, Zhong K, Li F, Zhang X, Yuan Z, Kang B, Degang X, Yao J. Temperature-dependent optical properties of $AgGaS_2$ in the terahertz range. Opt Mater. 2021;119: 111300.
10. Tell B, Kasper HM. Optical and electrical properties of AgGaS and $AgGaSe_2$. Phys Rev B. 1971;12(4):4455–9.
11. Boyd G, Kasper H, McFee J. Linear and nonlinear optical properties of $AgGaS_2$, $CuGaS_2$, and $CuInS_2$, and theory of the wedge technique for the measurement of nonlinear coefficients. IEEE J Sel Top Quantum Electron. 1971;12(7):563–73.
12. Harasaki A, Kato K. New data on the nonlinear optical constant, phase-matching, and optical damage of $AgGaS_2$. Jpn J Appl Phys. 1997;36(2):700–3.
13. McEwan KJ. High-power synchronously pumped $AgGaS_2$ optical parametric oscillator. Opt Lett. 1998;23(9):667–9.
14. Boon PP, Fen WR, Chong CT, Xi XB. Nanosecond $AgGaS_2$ optical parametric oscillator with more than 4 micron output. Jpn J Appl Phys. 1997;36(2):L1661–4.
15. Vodopyanov KL, Maffetone JP, Wieback IZ, Ruderman W. $AgGaS_2$ optical parametric oscillator continuously tunable from 3.9 to 11.3 μm. Appl Phys Lett. 1999;75:1204–1206.
16. Carnio BN, Zawilski KT, Schunemann PG, Elezzabi AY. Terahertz birefringence and absorption of a chalcopyrite $CdSiP_2$ crystal. Appl Phys Lett. 2017;111(22): 221103.
17. Carnio BN, Zawilski KT, Schunemann PG, Elezzabi AY. Optical rectification in a chalcopyrite $AgGaSe_2$ crystal for broadband terahertz radiation generation. Opt Lett. 2019;44(11):2867–3870.
18. Carnio BN, Greig SR, Firby CJ, Zawilski KT, Schunemann PG, Elezzabi AY. Terahertz electro-optic detection using a ⟨012⟩-cut chalcopyrite $ZnGeP_2$ crystal. Appl Phys Lett. 2016;108(26): 261109.
19. Łazewski J, Parlinski K. Lattice dynamics and elasticity of silver thiogallate ($AgGaS_2$) from ab initio calculations. J Chem Phy. 2001;114(15):46734–46738.
20. Wu BM, Chen ZJ, Zhu B, Du YL, He PS. Phonon softening of PTS single crystal during the phase transition at 200K. Chinese J Chem Phy. 1989;2(4):305–308.

21. Spitzer WG, Kleinman DA. Infrared lattice bands of quartz. Phys Rev. 1961;121(5):1324–35.
22. Barker AS Jr, Loudon R. Dielectric properties and optical phonons in $LiNbO_3$. Phys Rev. 1963;158(2):433–45.
23. Lavrentyev AA, Gabrelian BV, Ya Nikiforov I, Rehr JJ, Ankudinov AL. Ab initio calculations of the electron structure. J Phys Chem Solid. 2003;64(8):1251–5.
24. Holah GD, Webb JS, Montgomery H. Lattice dynamics of $AgGaS_2$. J Phys C Solid State Phys. 1974;7(21):3875–90.
25. Zhong K, Liu C, Wang MR, Shi J, Kang B, Yuan ZR, Li JN, Xu DG, Shi W, Yao JQ. Linear optical properties of $ZnGeP_2$ in the THz range. Opt Mater Express. 2017;7(10):3571–9.
26. Dmitriev VG, Lotsch HKV, Gurzadyan GG, et al. Handbook of nonlinear optical crystals;1999.
27. Vodopyanov KL. Optical THz-wave generation with periodically-inverted GaAs. Laser Photonics Rev. 2008;2(1–2):11–25.
28. Shi W, Ding YJ. Continuously tunable and coherent terahertz radiation by means of phase-matched difference-frequency generation in zinc germanium phosphide. Appl Phys Lett. 2003;83(5):848–50.
29. Shi W, Ding YJ, Schunemann PG. Coherent terahertz waves based on difference-frequency generation in an annealed zinc–germanium phosphide crystal: improvements on tuning ranges and peak powers. Opt Commun. 2004;233(1):183–9.
30. Creeden D, McCarthy JC, Ketteridge PA, et al. Compact fiber-pumped terahertz source based on difference frequency mixing in ZGP. IEEE J Sel Top Quantum Electron. 2007;13(3):732–7.
31. Rowley JD, Pierce JK, Brant AT, et al. Broadband terahertz pulse emission from $ZnGeP_2$ html. Opt Lett. 2012;37(5):788–90.
32. Rowley JD, Bas DA, Zawilski KT, et al. Terahertz emission from $ZnGeP_2$: phase-matching, intensity, and length scalability. J Opt Soc Am B. 2013;30(11):2882–8.
33. Fischer D W, Ohmer M C, Schunemann P G, et al. Direct measurement of $ZnGeP_2$ birefringence from 0.66 to 12.2 μm using polarized light interference. J Appl Phys. 1995;77(11): 5942–5945.
34. Zelmon DE, Hanning EA, Schunemann PG. Refractive-index measurements and Sellmeier coefficients for zinc germanium phosphide from 2 to 9 μm with implications for phase matching in optical frequency-conversion devices. J Opt Soc Am B. 2001;18(9):1307–10.
35. Bin K, Yunwei D, Mingjing T, et al. Growth of pure $ZnGeP_2$ crystals by horizontal gradient freeze method and its properties. J Chin Ceram Soc. 2016;44(4):503–7.

Chapter 5
Application of Terahertz Time-Domain Spectroscopy in the Field of Food and Agricultural Products

With the high-quality development of China's economy and significant improvements in people's living standards, the function of food has evolved beyond merely satisfying hunger, and its quality has gradually become a hot topic. Traditional food quality testing methods are convenient and cost-effective, but they have certain limitations or deficiencies. Modern common detection techniques include liquid chromatography-mass spectrometry, gas chromatography-mass spectrometry, liquid chromatography, gas chromatography, ultraviolet spectroscopy, infrared spectroscopy, and fluorescence spectroscopy, which can solve many problems that histology cannot [1, 2]. However, these methods have their limitations in practical applications, such as the large size of chromatography detection equipment, high testing costs, long testing time, and the need for professional personnel for preprocessing, resulting in low detection efficiency. In the field of food and agricultural products, there are usually a large variety and quantity of materials to be tested, which have a short shelf life and are not easy to store. Therefore, on-site real-time rapid detection is needed, and the above methods cannot meet the requirements for on-site rapid real-time detection of a large number of samples. In the process of realizing agricultural modernization, high-quality development of agriculture is a necessary requirement. Like food quality testing, the quality testing of agricultural products is also a major problem that needs to be solved urgently.

In recent years, many researchers have applied terahertz spectroscopy detection technology, a new type of non-destructive testing technology, to the field of food and agricultural product detection, and have achieved a series of important breakthroughs. Different chemical substances have different absorption characteristics for terahertz waves. By studying and analyzing the unique information of food and agricultural products in the terahertz band, different substances can be effectively identified and distinguished.

T. Chen et al., *Terahertz Time Domain Spectroscopy*,
https://doi.org/10.1007/978-981-95-2605-5_5

5.1 Detection of Food Additive Benzoic Acid

Supervising the use of food additives is very important for human health. Excessive use of food additives can harm the human body, especially potentially leading to organ failure or even cancer. Therefore, achieving a high-sensitivity detection of the widely used food additive benzoic acid is of great significance. Joint researchers from East China Jiaotong University and Zhejiang University, based on electromagnetic theory, designed a terahertz-enhanced metamaterial resonator and used the metamaterial resonator structure to enhance the detection of benzoic acid additives [3]. The absorption peak of this metamaterial resonator was designed at 1.95 THz, and the effectiveness of the metamaterial resonator was verified. The research results laid the foundation for the terahertz spectroscopic analysis of benzoic acid additives, enabling the terahertz technology detection of benzoic acid additives in food to meet the requirements of national standards. This research is of great significance for promoting the detection and analysis of trace additives in food, and its results can also provide a reference for the detection of antibiotic residues, banned additives, and other trace substances.

Firstly, a 500 μm quartz substrate was cleaned with acetone, methanol, and isopropanol in an ultrasonic bath for 10 min, and then baked on a hot plate for 2 min. Subsequently, the ultraviolet (UV) photoresist AZ5214 was spin-coated at 4000 rpm for 40 s, and baked at 90 °C for 6 min. Then, microstructures were drawn on AZ5214 using standard UV lithography techniques. After the development process, a 100 nm thick layer of gold was deposited on the sample using an electron beam evaporation tool. Finally, the metal pattern was obtained through a stripping process in an acetone ultrasonic bath for 20 min. Figure 5.1 shows the optical microscope image of the metamaterial structure.

Figure 5.2 presents the terahertz absorption coefficient spectra of benzoic acid solutions at different concentrations. The spectra were measured using the metamaterial. Considering that the front and back ends of the frequency range have more noise

Fig. 5.1 Optical micrograph of the prepared metamaterial structure

interference than other areas, the terahertz spectrum in the frequency range of 1.0–3.0 THz was selected as the research object for convenient subsequent data processing. As shown in Fig. 5.2, with the increase in the concentration of benzoic acid, the peak shows a trend of moving towards lower frequencies and decreasing. This may be due to the increase in surface reflection caused by the increase in concentration, thereby reducing the absorption peak. The metamaterial resonator has a strong absorption capacity at specific frequencies in the terahertz band and interacts strongly with the incident terahertz, forming a strong electric field. Compared with direct detection of the sample using terahertz waves, the interaction between the electric field generated by the metamaterial resonator and the sample is significantly enhanced, strengthening the sample information. After adding the sample to the surface of the metamaterial resonator, the surface dielectric environment of the metamaterial resonator changes, altering the impedance matching relationship between the air and the metamaterial resonator structure, causing a redshift in the resonance peak frequency. Therefore, the metamaterial resonator can enhance the interaction between the terahertz wave and the sample, and the frequency shift signal of the resonance peak in the metamaterial resonator can be used for quantitative detection and analysis of the sample.

Secondly, in order to eliminate the influence of factors such as uneven sample mixing, stray light, experimental environment, and sampling noise, SG (Savitzky-Golay) convolution smoothing, multiple scattering correction (MSC), and SG + 1st D preprocessing methods were chosen to preprocess the spectral data. The preprocessed spectral data and the content of benzoic acid in the aqueous solution were used as the input and output variables of the model, respectively, to establish a PLS model.

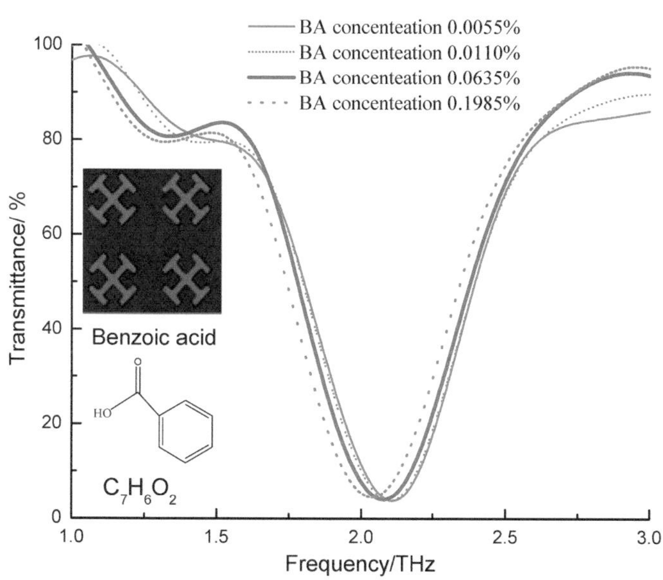

Fig. 5.2 Terahertz transmittance spectra at different benzoic acid concentrations

Then, the optimal preprocessing method was selected based on the Rp and a root mean square error of prediction (RMSEP) in the reference model. The results showed that the PLS model established after SG + 1st D preprocessing was the best, with Rp and RMSEP of the model being 0.9791 and 1.03×10^{-4}, respectively.

The optimal terahertz bandwidth selection method was explored using uninformative variable elimination (UVE) and competitive adaptive reweighted sampling (CARS). The wavelength variables selected by UVE and CARS were used as inputs, and the benzoic acid in the flour was used as the output to establish a PLS model.

Lastly, partial least squares (PLS) integrates multiple linear regression, principal component analysis, and canonical correlation analysis. When the spectral array and concentration array were decomposed simultaneously, their mutual relationship was also considered during the decomposition process. They also established a least squares support vector machine (LS-SVM) model, where the input was the wavelength variables selected by the UVE and CARS methods, and the variables were selected using the UVE and CARS methods combined with the principal component analysis method. Figure 5.3 shows the best prediction effect of the PLS model and LS-SVM model on the mixed sample benzoic acid concentration. For the PLS model, the wavelength selection of CARS not only reduces the calculation amount but also improves the accuracy of the model. When establishing the LS-SVM model, the wavelength selection using the CARS method can simplify the model and improve the accuracy of the model. Compared with the PLS model, the LS-SVM model has higher prediction accuracy. The prediction effect of the CARS-LS-SVM model established by the CARS wavelength selection is better, with a prediction correlation coefficient Rp of 0.9953, a RMSEP of 7.3×10^{-6}, and a detection limit of 2.3610 $\times 10^{-5}$ g/mL.

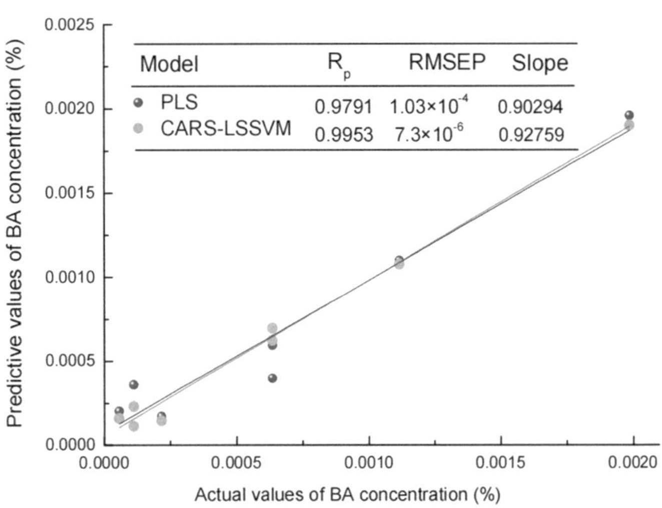

Fig. 5.3 Effect of PLS and LS-SVM models on optimal prediction values of benzoic acid concentration in mixed samples

5.2 Detection of Inverted Sugars in Honey—Food Additive

Adding cheap edible or non-edible ingredients can lead to serious health problems. Nevertheless, due to economic benefits, food additives are widely misused. Honey is a natural food that is popular worldwide, with an annual production exceeding 1.38 million tons. However, honey is easily adulterated with water, cane sugar, invert sugar, and syrup. The authenticity of honey is a major issue affecting consumer choice and market, damaging the interests of producers and disrupting the honey market. Fraudsters only need to add cheap sweeteners, such as sugar or syrup, to the honey to mimic the natural taste, thereby altering the characteristics of natural carbohydrates. Invert sugar syrup is a common adulterant in honey. The main components of honey are water and sugars. Water has strong absorption in the terahertz region. China Agricultural University used terahertz time-domain spectroscopy combined with attenuated total reflection (ATR) mode and chemometric methods for quantitative analysis of invert sugar honey [4]. In the experiment, a terahertz time-domain spectroscopy system (TAS7500SU) with an ATR module was used to measure the sample spectrum. A silicon crystal was used as the ATR prism, with a refractive index of 3.42 in the terahertz range. The terahertz pulse enters the ATR prism and then generates an evanescent wave at the interface of the ATR prism. The evanescent wave penetrates the sample, so the power difference of the reflected evanescent wave contains the information of the sample. The air and honey on the ATR prism were scanned 1024 times, and then the average spectra were taken as the reference and sample, respectively. 500 uL of honey sample was pipetted onto the ATR prism. The sample flowed slowly under the action of gravity and then remained stationary during the measurement. It is necessary to clear the interface gap between the ATR prism and the honey to reduce the effects of reflection and scattering.

The spectra of different invert sugar syrup percentages from 10 to 90% (w/w) differ in the slope of the curve. The proportions of water in acacia honey and invert sugar syrup are 16.16% (w/w) and 20.90% (w/w), respectively. As the proportion of invert sugar syrup increases, more water molecules replace sugar molecules, and the absorption rate of sugar molecules in the terahertz band is lower than that of water. In addition, the measurement of sugar solution is mainly driven by the intermolecular vibration mode produced by the hydrogen bond interaction between water and sugar, rather than the intramolecular mode in the terahertz band. Adding more invert sugar syrup can enhance the interaction of hydrogen bonds, which is ultimately stronger than van der Waals forces. Although the absorption spectra overlap in some frequency ranges, the absorption spectra corresponding to each ratio can be well distinguished. At the same time, there is a certain rule between the absorption coefficient and the proportion of invert sugar syrup. Therefore, the linear correlation between the spectral characteristics and the adulteration ratio is discussed next.

The increasing trend of the effective frequency area in the absorption coefficient spectrum is related to the concentration of invert sugar syrup. Therefore, using the SLR model quantifying the proportion of invert sugar syrup, the best fitting curve is found on the two-dimensional plane of the absorption coefficient under the full terahertz spectrum. At the same time, they used the SLR model to test the relationship between the derivative of the absorption coefficient spectrum and the percentage composition of invert sugar syrup at different bands. Using the statistical SLR method, an evaluation model was established between the adulteration percentage content and the comprehensive absorption coefficient spectrum from 0.5 THz to different frequencies of acacia honey adulterated with different proportions of invert sugar syrup (i.e., the area of the absorption coefficient spectrum).

Since there is no specific absorption peak in the terahertz band, it is necessary to discuss different SLR models at different frequencies. Therefore, these three spectral characteristics and different SLR models between the proportion of invert sugar syrup were compared with different bands.

Different SLR models at different bands will have different results. The adjusted R^2 value increases first and then decreases with the increase of frequency, while the RMSE value decreases first and then relatively increases. For the SLR model between the absorption coefficient and the proportion of invert sugar syrup, there is a good correlation in the frequency range of 1.5 THz to 3.5 THz, with a adjusted $R^2 > 0.9$, RMSE < 5. However, in the frequency range below 1.5 THz and above 3.5 THz, there is a lower adjusted R^2 value and a higher RMSE, so there is no good correlation. It should be noted that the unstable fuctuations ofthe adjusted R^2 and RMSE indicate the instability and error-proneness of the model. As for the other two spectral features, the SLR model has a slight dependency on the selected bands. Although within the frequency range of 1.2–3.2 THz, the adjusted R^2 value of the area model is less than that of the slope model, and the RMSE of the area model is greater than that of the slope model, in most of the wavelength range, the spectral slope and the conversion syrup ratio show a good linear relationship, especially in the range where the frequency is less than 1.2 THz and greater than 3.2 THz.

Therefore, a method was proposed to analyze the dependency of the conversion syrup ratio from three spectral features, and its optimal regression equation was established. The deviation of the adjusted R^2 value of the entire selected model from 1 is less than 0.02, indicating that the selected model has a good linear relationship. All regression coefficients' F-tests have significant effects ($P < 0.001$).

To compare these three spectral features, they chose three better SLR models for further analysis. The research results show that the SLR model may be a valuable model for quickly identifying conversion syrup. It can eliminate the multivariate statistical calculations and complex evaluation models of quantifying the conversion sugar ratio. They reported the application of the SLR model in terahertz spectroscopy technology. The independent variable is usually the absorption coefficient at a certain frequency. But for complex samples, due to the wide absorption spectrum, it is difficult to establish a model using a determined absorption coefficient. Using the area of the absorption coefficient spectrum is more suitable for detecting complex samples. Therefore, the SLR model of the absorption coefficient area is also a reasonable quantitative detection method.

The SLR model only uses one variable, so it may cause the loss of valuable information during the research process. If the independent variable has noise characteristics, then the prediction of the dependent variable is easily affected by changes.

Therefore, they used the MLR model to combine several information variables for multivariate calibration. MLR is a consensus strategy of a representative subset of variables with minimal collinearity, in most cases selected by SPA. There are a total of 657 original frequencies, and based on SPA, partial frequencies are selected for the absorption coefficient model, slope model, and area model. After variable selection, three corresponding MLR equations were established. After the F-test, these three MLR models have significant effects ($P < 0.001$). There are no systematic errors in the results, and the points of the calibration set and validation set are randomly distributed around the diagonal. All selected variable MLR models are acceptable (prediction bias ratio (RPD) > 2). The frequency obtained from SPA can represent the main spectral information, which is an effective method of variable selection. At the same time, the MLR model with the absorption coefficient spectral slope is slightly better than the MLR model with the absorption coefficient. The MLR model with the absorption coefficient spectral area performs better, but performs poorly in the calibration set. The area model used has 12 variables, making the model simple enough and having the best RPD (>5), indicating that the model has good predictive ability. Among the above three models, the area model's prediction results for the validation set are better than other models. Therefore, an MLR model with an absorption coefficient spectral area can be chosen for construction in applications. Despite this, further research is still needed on the application of the MLR model using the absorption coefficient area for other types of samples. In addition, it is worth emphasizing that the calibration set and validation set results of the MLR model with the absorption coefficient spectral area are better than the SLR model.

In conclusion, the research results show that terahertz spectroscopy combined with the MLR model can reliably and accurately assess the conversion syrup concentration ratio in acacia honey. Considering the practical application of variable selection, whether using the SLR model or the MLR model, compared with using the absorption coefficient and absorption coefficient spectral slope, modeling using the area of the absorption coefficient can improve accuracy and robustness. Linear models with fewer parameters are simpler and easier to use than nonlinear models; at the same time, linear models have stronger capabilities to assess adulteration. Therefore, linear models, especially MLR models with SPA, will be used more frequently for real-time and on-site quantitative assessment of adulteration in honey.

5.3 Sunflower Seed Breeding Detection

Over the past few decades, sunflower seed oil extracted from sunflower fruits has become popular worldwide. The fullness of sunflower seeds is a key indicator reflecting seed vitality, which has a significant impact on seed yield and oil quality. Generally speaking, seeds that are large and undamaged have high vitality, and planting them can easily yield high-quality and high-yield results, that is, fuller seeds can improve the seed load and quality of each plant. Therefore, fullness measurement helps to clarify its response characteristics in the terahertz field, providing technical support for seed quality control. In 2020, X. D. Sun and J. B. Liu from East China Jiaotong University conducted a study on the fullness of sunflower seeds using terahertz transmission imaging technology [5]. To fully investigate the performance of terahertz in thin-shell seed detection, the samples used included sunflower seeds that had been nibbled by worms, incomplete sunflower seeds, and complete sunflower seeds. The thickness of healthy sunflower seeds is between 3–5.5 mm, the appropriate moisture content is below 10%-12%, and the moisture content of the samples used in this experiment is less than 80%, which can reduce the interference of water on the absorption of terahertz waves. A terahertz time-domain transmission imaging system (TAS7500TS, Advantest) was used, as shown in Fig. 5.4. After the terahertz images were collected, the shell and kernel were carefully separated by manual operation. To simplify the image process, the sunflower seed shells and kernels were photographed separately against a pure black background, using an RGB camera at the same height, short exposure time, and the same light intensity.

The terahertz images extracted using Matlab2013 software were subjected to manual threshold judgment, realizing the segmentation of complete seeds and background as well as seed shells and kernels, and the fullness was calculated by the area ratio of kernels and shells, expressed in %. Since the edge image is too thin, it may be lost during the terahertz imaging and segmentation process. Therefore, after the terahertz images were collected, the shells and kernels of the sunflower seeds were manually separated to maintain their integrity. Next, their RGB images were collected. Using a manual re-inspection method, samples with incomplete kernels and worm damage were confirmed for qualitative identification. The sunflower seed

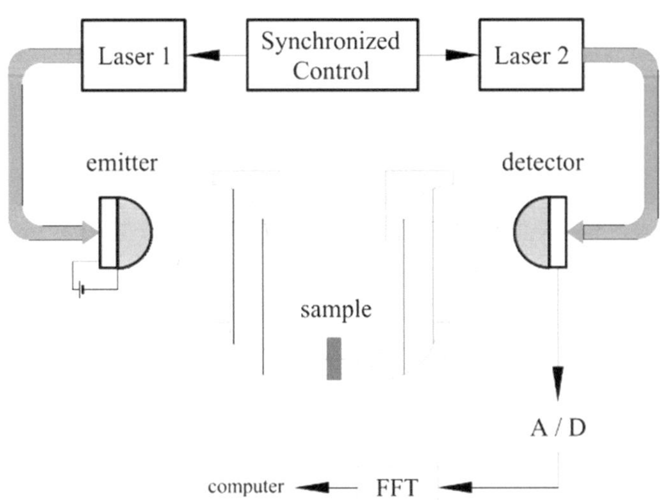

Fig. 5.4 Schematic diagram of terahertz transmission imaging experimental setup

RGB images were threshold segmented, and the Otsu method was used to automatically judge the threshold, a method of adaptive threshold judgment proposed by Japanese scholar Nobuyuki Otsu [6]. Similar to the exhaustive method, the image is divided into the background and the region of interest (ROI). The larger the inter-class variance, the better the segmentation effect. After obtaining the segmented images of the sunflower seeds, the fullness was calculated using the RGB images as a reference value for calibrating the terahertz image fullness measurement model.

First, the terahertz spectral characteristics of sunflower seeds were analyzed. The terahertz images of sunflower seeds were collected using a terahertz imaging system, and the shell and kernel were observed with the eye, and the region of interest was manually selected to extract the time-domain signal and absorption coefficient. To eliminate random noise, the team extracted the average spectrum of 3×3 pixels in the shell and kernel areas of each sunflower seed image. The time-domain signals of the background (PE material), sunflower seed kernels, and shells are shown in Fig. 5.5a, b shows the absorption coefficients of the corresponding samples obtained after Fourier transformation using the TAS7500TS system from Advantest. The main components of the sunflower seed shell are cellulose and starch, accounting for 52.16% and 26.48%, respectively. In the frequency range of 0.5–3.5 THz, the absorption capacity of cellulose for terahertz waves gradually increases. Due to the uneven thickness and density distribution of cellulose, and the low content of free water in cellulose, the spectral intensity of the sample may vary slightly. The cellulose content in the seed shell is high, and its absorption capacity in the 0.5–3.5 THz frequency range is strong. The high molecular weight starch mixture has no obvious absorption peak in the 0.5–3.5 THz frequency range, but its absorption gradually increases in the 0.5–2.0 THz range. Sunflower seed kernels are rich in fats and proteins, accounting for 53.32% and 29.92%, respectively. The absorption intensity of sunflower seed kernels in the frequency range of 0.5–2.5 THz increases with the increase of frequency, and the absorption of fats and proteins is strong, and the amplitude of the time-domain signal is almost zero. There are obvious differences in the absorption characteristics of sunflower seed shells and kernels in the frequency range of 0.5–2.0 THz. At the same time, the team studied the attenuation of terahertz radiation for sunflower seeds with moisture contents of 5%, 10%, and 15%, and the influence of moisture content on the grayscale of terahertz images is small.

Next, the fullness of sunflower seed kernels based on terahertz images was studied. The area ratio of kernels and shells was calculated using the terahertz images of complete sunflower seeds. To solve the edge problem in image feature extraction, the area ratio of the RGB images of the peeled seed shells and kernels was used to correct and establish a prediction model for the fullness of sunflower seeds in terahertz images. The spatial resolution of the terahertz images collected in their experiments was 38×63. Since there is some noise in the background of the original terahertz image, the pixel value σ is less than 0 in the background. To make the image background clear, they set this pixel value to 0. Each column in the pixel matrix was summed, and according to the statistical results, the column with the sum of pixel values as $\sigma = 0$ was used as the separation boundary, dividing the four sunflower seeds in each image into four independent image areas. The threshold was used to

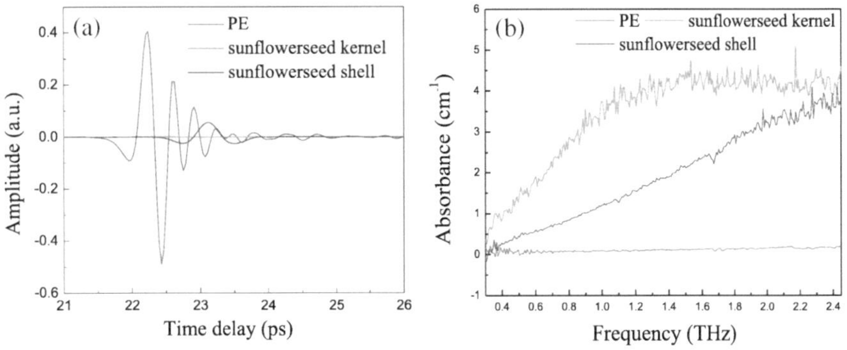

Fig. 5.5 Terahertz time-domain spectrum **a** and frequency-domain spectrum **b** of PE, sunflower seeds, and shells

judge the shell and core areas in the image. $T_s = 0$ was taken as the threshold of the seed shell, when $\sigma > T_s$, the pixel represents the sunflower seed shell area, and its area S_{T1}. Due to the low resolution of terahertz images, the threshold for the sunflower seed shell region needs to be manually determined. The pixel values of the sunflower seed grain region and the shell region are significantly different, making it easy to determine T_k. When $\sigma > T_k$, the pixel represents the seed kernel, and its area S_{T2} can be calculated by counting the number of pixels. The fullness of the sunflower seed terahertz image is P_T:

$$p_t = \frac{S_{T2}}{S_{T1}} \times 100\% \tag{5.1}$$

where P_T represents the fullness of the sunflower seed terahertz image, and S_{T1} and S_{T2} represent the number of pixels in the seed grain and shell regions of the image, respectively.

Sunflowers may suffer from plant diseases or insufficient growth, leading to internal defects such as wormholes or seed grain tearing. These internal defects are difficult to distinguish with the eye and greatly affect their usability and seed quality, thus need to be removed. The composition and content of cellulose, protein, and fats in sunflower seed shells and grains show different absorption intensities to terahertz waves. These defects will affect the spectral characteristics of terahertz waves, creating gaps and defects in terahertz images, and identifying seed shells from a terahertz perspective.

As shown in Fig. 5.6, dry, worm-eaten, empty, and broken seed samples were mixed with intact seed samples and imaged using a terahertz time-domain spectroscopy system. From the image, it can be seen that the seed grain area of dry, worm-eaten, empty, and broken seed samples contains almost no protein, fats, etc., and the terahertz absorption is weak, as can be seen from the results of Fig. 5.6a–d. From Fig. 5.6h, it can be seen that the terahertz image of the defective sunflower seed sample is clear and consistent with the RGB image.

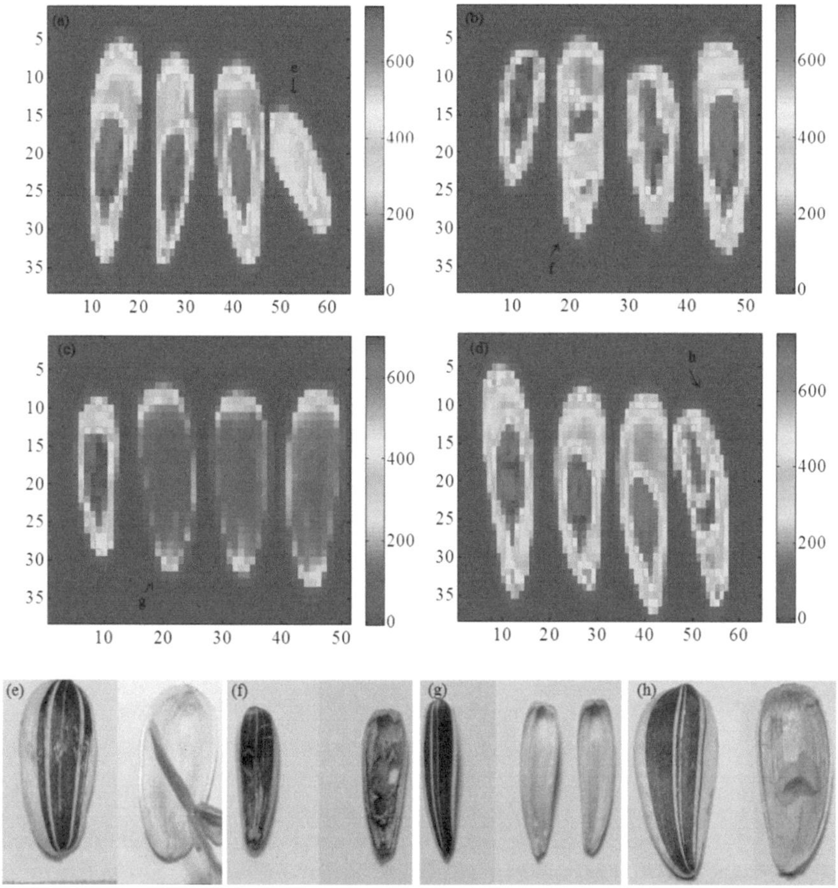

Fig. 5.6 Terahertz imaging results of different samples

To better judge the fullness of the seeds, it is necessary to further qualitatively analyze the broken seeds and worm-eaten seeds and other incomplete sunflower seeds. According to the related RGB images, 80 groups of sunflower seeds were manually divided into perfect and imperfect groups based on the brightness of the terahertz images. By studying the fullness of the seeds and calculating the average values of these two groups, the average fullness of the perfect grains is $P_s = 34.7\%$, and the average fullness of the imperfect grains is $P_u = 6.4\%$, where the fullness threshold $T_p = (P_s + P_u)/2 = 20.6\%$, as shown in Fig. 5.7. Whether a sunflower seed is defective can be determined by the seed fullness threshold T_p. When the terahertz fullness $P_T < T_p$, it is an imperfect seed; when $P_T > T_p$, it is a perfect seed. According to the fullness threshold, the recognition rate of the terahertz imaging system for incomplete seeds is 97%. The structure of this study is consistent with previous

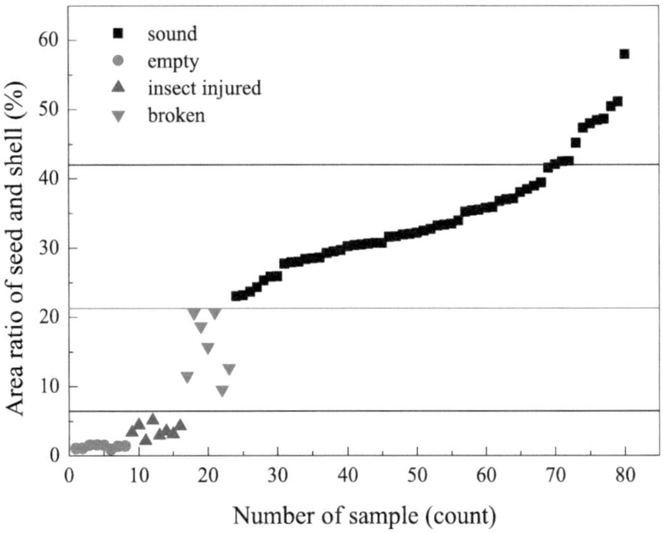

Fig. 5.7 Saturation of different samples

reports on the use of terahertz time-domain spectroscopy in identifying plant seed varieties or detecting defects.

They also adjusted the terahertz fullness calculation model. The fullness of complete sunflower seeds was selected for modeling, and data from 57 good samples were obtained, and imperfect sunflower seeds were removed from the terahertz image fullness data. The terahertz fullness data was divided into a calibration set and a prediction set at a ratio of 43:14. The model was established between terahertz and RGB images, and the determination coefficient of calibration (R_c^2) and the root mean square value of calibration were 0.9 and 3%, respectively. In the sunflower seed fullness experiment, the fullness was measured using a terahertz imaging system, and the selected prediction set ($n = 14$) was used to replace the model function for model verification. The prediction results and actual values have a good linear relationship, the determination coefficient (R_p^2) is 0.91, and the root mean square error of prediction is 4%, as shown in Fig. 5.8.

5.4 Non-Destructive Detection of Cigarette Burst Beads

Capsules filled with flavoring solution are embedded in cigarette filters. These flavor capsules contain liquid flavorings that help enhance the taste of the cigarette. The capsules are embedded in the cigarette filter, which is marked on the surface and can be crushed by the consumer during use. The aromatic liquid is then released from the capsule, allowing the user to experience a specific flavor. Capsules containing liquid additives aim to make cigarettes more delicious and create a desirable sensory

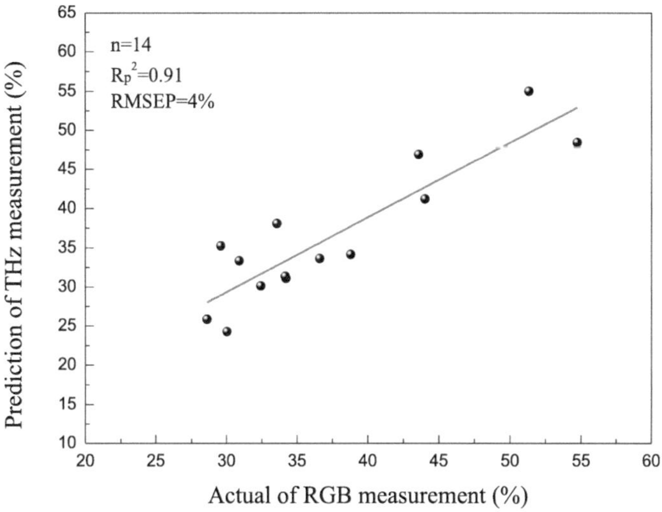

Fig. 5.8 Prediction model for sunflower seed fullness

experience. However, on the cigarette production line, there are some quality issues with the capsules, such as displacement with the movement of the mark, missing capsules, and leakage of aromatic liquid. Online detection of tobacco filters and capsule positions plays an important role in tobacco quality control. Many methods have been used for quality control of tobacco filter capsules. Using a strong light box exposure method to locate the capsules in the filter is harmful to the operator's eyes and has low efficiency. Microwave technology can be used to find the position of the capsule, but the corresponding resolution is low. X-ray imaging is also a detection method, but due to its radiation and environmental protection assessment regulations, this method is not an ideal choice for engineering implementation. Terahertz waves can penetrate tobacco filters and locate capsules with high resolution and non-ionizing radiation. The diameter of the filter capsule is between 2.6 and 7.6 mm, which matches the spatial resolution of terahertz waves. In addition, due to its high transmittance and high resolution, terahertz waves have broad application prospects in the quality control of tobacco filter capsules. The authors proposed a method for non-destructive testing of tobacco filter capsules using terahertz time-domain spectroscopy [7]. In the experiment, a transmission terahertz time-domain spectroscopy imaging system was used to measure tobacco filter capsules. The results show that through terahertz transmission imaging, the dual capsules in the tobacco filter can be located and distinguished. Terahertz technology has good application prospects in improving the quality control of tobacco filter capsules and improving production efficiency.

An asynchronous optical sampling THz-TDS system was used as the detection equipment. Unlike traditional terahertz time-domain spectroscopy systems, this system does not require a mechanical delay stage. Two femtosecond lasers are synchronized by an electronic control board. The wavelengths and pulse bandwidths of the two lasers are the same. However, the repetition rates of the two lasers are slightly different. The pulsed femtosecond laser irradiates the photoconductive antenna transmitter and detector, then generates pulsed terahertz waves and performs corresponding detection. Compared with traditional terahertz time-domain spectroscopy technology, asynchronous optical sampling is faster because there is no mechanical delay platform. In addition, electronic control sampling is more stable than mechanical methods.

Tobacco samples were collected from the consumer market. In the cigarette filter, there are two filter capsules. Between the cigarette filter paper and the pipe, the filter package paper is printed with a metal foil tag marked “555”. The tobacco sample is fixed on the sample rack.

Terahertz spectra and imaging results were obtained in the experiment to detect filter capsules. The filter capsules contain liquid flavorings that can absorb terahertz waves. The terahertz signal of the capsule is much weaker than the terahertz signal of the tobacco filter area without the capsule, and there is a high terahertz response contrast between the two waveforms.

To intuitively display the position of the filter capsules, the cigarette sample was scanned using an XY two-dimensional platform, with a spatial resolution set to 200 μm. The positions of the two capsules in the tobacco filter can be clearly displayed. Since the brand mark “555” is made of metal material, part of the terahertz signal is blocked. In addition to the two obvious filter capsule points, there is low contrast point on the left side of the capsule.

5.5 Detection of Capsaicin in Soybean Oil

Recycled cooking oil, also known as “gutter oil”, refers to oil products that are illegally processed and resold as cheap cooking oil. If qualified vegetable oil is mixed with gutter oil or other discarded cooking oils, it can harm the human body, causing abdominal pain or diarrhea. Therefore, researching a direct and sensitive method to identify recycled cooking oil has practical application value. Capsaicin is the main irritating component in chili peppers, with capsaicin and dihydrocapsaicin accounting for about 90%. Frying, cooking, or using chili peppers and pepper-containing seasonings in food will cause capsaicin to steadily dissolve in the cooking oil. Chili oil has lipophilicity and stability in the cooking and refining process of waste cooking oil, so capsaicin is inevitable and remains in the used oil after the chili oil frying process. Capsaicin is a key component of chili peppers and can serve as a marker for gutter oil. The research group of Changhong Liu from Hefei University of Technology studied the feasibility of rapidly detecting capsaicin concentration in soybean oil based on terahertz spectroscopy technology [8]. Using genetic algorithms (GA) and principal component analysis (PCA) as preprocessing methods, they obtained the optimal spectral features. By combining the least squares support vector machine (LS-SVM), backpropagation neural network (BPNN), and partial least squares neural network (PLS) with preprocessing methods, they obtained the optimal determination model. By combining BPNN and genetic algorithms, they obtained the best quantitative prediction results, with prediction correlation coefficient (Rp), prediction root mean square error (RMSEP), prediction bias ratio (RPD), and range error ratio (RER) of 0.9309, 0.4030ug/kg, 17.0421, and 2.4813, respectively. In addition, after preprocessing with LS-SVM combined with GA, the detection limit of capsaicin in soybean oil can reach 1.25 μg/kg, with an identification accuracy rate of up to 100%. The results show that terahertz spectroscopy combined with chemometrics has great potential in rapidly determining the concentration of capsaicin in soybean oil.

Capsaicin stock solution was injected into pure soybean oil to obtain soybean oil samples with different concentrations ranging from 0 to 20 μg/kg. Each capsaicin concentration soybean oil sample was divided into 100 samples to obtain terahertz spectral information. Therefore, they obtained six different concentrations of capsaicin, a total of 600 samples, and randomly divided them into a calibration set (80×6 samples) and a prediction set (20×6 samples). The soybean oil samples with different capsaicin concentrations were classified, with accuracy (A%) as the model selection criterion. For the quantitative prediction of capsaicin concentration, the model quality was evaluated using calibration root mean square error (RMSEC), RMSEP, calibration correlation coefficient (Rc), Rp, RPD, and RER.

The samples were measured in transmission mode using the TAS7500TS system. In the time-domain spectrum, the transient electric field of the terahertz wave is a pulse signal, and different samples have differences in amplitude and phase field. This may be due to the differences in capsaicin concentration in the samples. Since the terahertz frequency range is wide, it is not clear in both amplitude and phase field. Due to the limitations of the instrument, only information within the frequency range of 0.5–2.5 THz is given here. The difference in the absorption spectrum is more obvious than the time-domain spectrum and frequency-domain spectrum. Due to the high resolution of the terahertz spectrum, the dimension of the terahertz absorption signal data exceeds 300. Because the original data of the terahertz spectrum is rich in information, it is difficult to obtain a stable and accurate determination model with a simple method. Therefore, they used chemometric methods such as LS-SVM, BPNN, PLS, combined with PCA, GA, etc., to obtain the determination model of capsaicin in soybean oil.

First, a principal component analysis was performed to observe the qualitative differences in different capsaicin concentrations in the principal component space. The absorption spectrum is the most suitable feature for detecting different capsaicin concentrations in soybean oil samples. The spectral changes of the first three principal components are 30.9%.

Genetic algorithms and principal component analysis are used as preprocessing methods for terahertz spectral data, reducing high-dimensional information to obtain optimal features. Concurrently, BPNN, LS-SVM, and PLS are combined with preprocessing methods to obtain the optimal model for predicting and classifying capsaicin concentration in soybean oil samples. The results show that, except for the GA-PLS model, the correlation coefficients of the quantitative prediction models are all above 0.9. For different prediction models, RMSEC, RMSEP, RPD, and RER vary significantly. When quantitatively predicting different concentrations of capsaicin in soybean oil, BPNN combined with GA is the optimal model. Although the correlation coefficient of BPNN combined with PCA is high, the root mean square error is high, and RPD and RER are low, indicating poor predictive performance. Using BPNN and PLS models for prediction, the prediction results without data preprocessing are more accurate than those with data preprocessing. Compared with BPNN and PLS models, the LS-SVM combined with GA preprocessing has the best discrimination ability, with an accuracy of 100% on both the calibration set and the prediction set. Moreover, terahertz spectroscopy technology can successfully detect a capsaicin concentration of 1.25 μg/kg in soybean oil, with a prediction accuracy of 100%. It can be seen that terahertz spectroscopy technology combined with chemometrics has good application potential in the measurement of capsaicin concentration and the identification of recycled edible oil.

5.6 Detection of Different Sources, Varieties, and Transgenic Populations

The geographical origin of plant-based agricultural products is crucial for ensuring the authenticity and traceability of mislabeled and fraudulent products. The quality and composition of products from different sources will vary. Terahertz spectroscopy technology can non-destructively detect the same type of samples from different sources.

Corn is a globally important economic crop, consumed as a staple food by most of the world's population and used as a basic feed for livestock. High-oil corn (HOC) is an attractive new corn variety with an oil concentration of 6%, twice that of regular corn. In addition to generating more heat energy with such a high oil content, high-oil corn also contains high-quality protein, making it an ideal choice for the food, feed, and bio-refining industries. Therefore, controlling the purity and authenticity of high-oil corn is one of the current priorities in agricultural research, with corn quality testing and corn variety identification playing a central role in these research activities. Researchers from several universities, including Tianjin University, used terahertz spectroscopy and signal classification analysis to identify corn varieties [9]. Features and variable selection from the original signals of terahertz time-domain waveforms (THz-TDW) and absorption spectra (THz-AS) were extracted, and classifiers were used on these processed signals to establish identification models. Principal component analysis (PCA) was performed on THz-TDW for feature extraction, and three different variable selection methods were implemented on THz-AS, including uninformative variable elimination (UVE), uninformative variable elimination successive projection algorithm (UVE-SPA), and competitive adaptive reweighted sampling (CARS). Then, two classification algorithms, linear discriminant analysis (LDA) and support vector machine (SVM), were used and compared in the discrimination models. These models were validated using the bootstrap Latin partition (BLP) method with 10 bootstrap partitions and 5 Latin partitions. Their model results show that SVM is a better classification algorithm with higher recognition accuracy, and the accuracy of the PCA-SVM model of THz-TDW reached 94.7%. Variable selection is an important step in creating an accurate and robust THz-AS recognition model. The CARS-SVM model with a radial basis function (RBF) achieved an average accuracy of 100% on the prediction set, while the accuracies of the UVE-SVM and UVESPA-SVM models were 91.2% and 99.1%, respectively. The results show that terahertz spectroscopy and discriminant analysis can successfully identify high-oil corn and regular corn, providing a reference for the classification of agricultural products in agricultural research, and providing an effective reference for expanding the application of terahertz spectroscopy in related fields.

In the study, two corn varieties, high-oil 5598 (HOC, $n_{HOC} = 360$) and Zhengdan958 (NC, $n_{NC} = 210$), were analyzed. Typically, high-oil corn (HOC) has an oil content of 6% to 10%, while regular corn (NC) has an oil content of 3% to 6%. Each corn sample was taken from the middle area of the corn cob, first ground into powder with a grinder, then ground into fine powder with an agate mortar, and finally pressed into smooth small pieces to eliminate the impact of multiple scattering. Finally, 0.2 g of corn powder was used to make tablet samples, with the pressing conditions being a pressure of 15 t for 15 min.

The terahertz time-domain waveforms of different corn varieties show slight differences, indicating similar compositions, but different components and refractive indices. Since corn is composed of various components such as starch, fat, protein, and various amino acids, there are no obvious characteristic absorption peaks in the terahertz absorption spectrum. The absorption of terahertz waves by high-oil corn and normal corn increases with frequency, and within the measurement range, the absorption of normal corn is higher than that of high-oil corn. The absorption of terahertz waves is mainly related to the polar molecules (such as amino acids, starch, and protein) in the corn samples, so the decrease in oil content and the increase in starch content in normal corn lead to an increase in terahertz absorption. In addition, in the frequency range of 1.5–2.5 THz, the difference between high-oil corn and normal corn is much greater than other frequency ranges. Terahertz absorption spectroscopy is more intuitive when analyzing the properties of materials. On the other hand, the terahertz time-domain signal spectrum provides signal characteristics related to the absorption and refractive index of the sample. Both are important indicators for corn identification research.

Results of principal component analysis of terahertz time-domain signals. Extracted from the terahertz time-domain signal spectrum through PCA, the first principal component (PC1) captured 72.33% of the total variance, while the second principal component (PC2) and the third principal component (PC3) captured 25.87% and 1.44% of the total variance, respectively. Since the cumulative contribution rate of these three principal components has reached 99.64%, they contain most of the signal variance information, while other principal components are mainly related to noise variance. The scores of PC1, PC2, and PC3 of the terahertz time-domain signal spectrum were also studied. The data were divided into two different clusters related to high-oil corn and normal corn samples. These two clusters overlap slightly, which is caused by the strong multicollinearity between the terahertz time-domain signal characteristics of corn.

Identification results of terahertz time-domain signals processed by PCA. In order to establish an identification model, they used the first three principal component factors of the terahertz time-domain signal as input for LDA and SVM classifiers. Compared with directly using the original terahertz time-domain signal waveform as input, after PCA processing, the prediction accuracy of all classification models has been improved. The LDA model achieved an accuracy rate of 77% in prediction testing, while the accuracy rate of the SVM model using RBF reached 94%. In addition, the SVM with RBF has the highest sensitivity because it has many unique advantages in small sample areas and nonlinear and high-dimensional pattern recognition compared to LDA. It also has the narrowest confidence interval, proving the credibility of the model.

Identification results using terahertz absorption spectra. Subsequently, they applied UVE, UVE-SPA, and CARS to select variables from the original terahertz absorption results, and then input them into the identification model. In this study, UVE selected 254 variables from the original 393 variables. The selected variables account for 64.6% of the initial variables, but do not meet the expected selection efficiency of corn terahertz absorption information variables. Since UVE retains a large number of related variables, SPA is applied to further reduce the number of selected variables. The increase in feature variables can improve the accuracy of prediction. In order to obtain the most informative variables, the F-test criterion is used to determine the optimal number of variables. UVE-SPA finally selected 33 feature variables. Compared with UVE alone, further variable selection using UVE-SPA is more conducive to establishing a robust identification model.

With the increase in sampling times, the number of sampling variables decreases rapidly first, and then decreases slowly, indicating that the two-stage selection of CARS is fast and refined selection. The accuracy slowly improves in the 1–35 sampling runs, which should be due to the removal of variables without information. When the sampling times exceed 35, the accuracy begins to decline. Based on the optimal sampling run using the CARS algorithm, 64 feature variables were selected.

These information variables are related to the characteristic absorption frequencies of the terahertz absorption spectra of high-oil corn and normal corn. Most of the variables selected by the three methods (UVE, UVE-SPA, and CARS) are distributed in the frequency ranges of 0.5–1.5 THz and 1.5–2.5 THz. Fewer variables are selected in the 2.5–3.5 THz frequency range because the absorption coefficients of high-oil corn and normal corn in this area are similar. Therefore, the selection of variables will be affected by the low signal-to-noise ratio caused by high absorption in the 2.5–3.5 THz range.

Identification results of terahertz absorption spectra in different frequency ranges. Compared to the full frequency band, range division can enhance the accuracy of the discrimination model. For the terahertz absorption spectrum within the frequency range of 1.5–2.5 THz, the average accuracy of the LDA, SVM models with LF and RBF in the prediction set are 83.3%, 87.7%, and 90.4% respectively. The lower discrimination accuracy within the 2.5–3.5 THz frequency range may lead to redundant and unusable variables, as well as a lower signal-to-noise ratio. Compared to the LDA model, the SVM model has lower specificity in the prediction set. However, the prediction accuracy of the SVM model is superior to the LDA model. For the SVM model, using RBF has a higher recognition rate than LF.

Discrimination results of terahertz absorption spectra with variable selection methods. The sensitivity, specificity, and accuracy of discrimination models using different classifiers SVM and LDA for three variable selection methods were studied. In comparison, the SVM model always achieves better performance and narrower confidence intervals than the LDA model. Among the three variable selection methods, the CARS-SVM model with RBF has an average accuracy of 100% in the prediction set, while the accuracy of the UVE-SVM and UVE-SPA-SVM models are 91.2% and 99.1% respectively. Due to the high non-linear pattern recognition ability of the SVM model, which finds the optimal discrimination boundary through mapping from low-dimensional space to high-dimensional space, it has higher sensitivity and specificity. Compared to the full frequency band, the variable extraction method reduces the confidence interval of the SVM and LDA models in the validation set.

By selectively using feature variables, the accuracy of the recognition model is significantly improved compared to using the original terahertz absorption spectrum. Among these variable selection methods, CARS is the most effective for both LDA and SVM models. By applying adaptive reweighted sampling and exponential decay functions, CARS can effectively eliminate variables that contribute minimally to the model. The specificity of UVE is relatively low because it only removes variables closest to noise variables, ignoring other irrelevant information that needs to be rejected. By first removing noise variables with UVE, UVE-SPA obtains variables with the smallest linear relationship, which can establish the most simplified model, but may also lead to over-removal of variables. However, compared to CARS (64 variables) and UVE (33 variables), UVE-SPA has the fewest feature variables (254 variables).

5.7 Fungal Detection in Chestnuts

The quality of chestnuts is related to characteristics such as size, shape, color, taste, and texture, which are very important for assessing the freshness and edibility of chestnuts and the processing process. Chestnut fruits contain sugars, mainly monosaccharides and disaccharides. Due to this organic structure, chestnuts are very attractive to insects, especially fungi, which can cause serious damage to chestnut cultivation. In addition, because food quality and human health are directly related, real-time safety monitoring of food is necessary.

In some European regions (such as Italy, France, and Switzerland), Gnomoniopsis castanea is a new type of fruit rot fungus. G. castanea in chestnuts initially causes a change in the color of the endosperm, followed by mummification in the colony area [10]. In addition, fungi such as Penicillium, Aspergillus, Fusarium, Coniothyrium, and Pseudomonas were also found in chestnuts. Timely individualization of pathogens in the case of infection is an urgent issue. However, non-destructive methods that can detect infections in living tissues in real-time and at an early stage are still lacking. The biodegradation of chestnut fruits can be caused by the presence of fungi, which may be related to changes in the transmittance of the fruit. In addition, the chestnut shell is thin and basically dry, so it is basically transparent in the terahertz band, which allows measurements to be made without opening the chestnut. In 2021, Di Girolamo et al. in Italy used a continuous terahertz wave transmission imaging system to determine whether there is a fungal infection in biological samples (i.e., chestnuts) of a few millimeters in size by detecting differences in quality or volume attenuation density, without the need for sample pretreatment [11].

The chestnut material samples used in the experiment were taken from Teano (Italy). All chestnuts were placed in plastic bags and then kept in a refrigerator at + 8°C until measurement. Each chestnut was dissected into two parts. The exposed inner surface was divided into four quadrants, and a small part of the tissue was cut from each quadrant, two parts of tissue were cut from half of the fruit sample, two parts were cut from another, and then spread on potato dextrose agar.

The terahertz imaging system consists of an Impact Avalanche Transit Time (IMPATT) diode source (IMPATT-100-H/F, output at 100 GHz is 80mW, Terasense) and a terahertz camera based on a sensor array (Tera-1024 The setup consists of a 32 × 32 Terasense array. Lower signals detected on the detector correspond to higher terahertz intensity attenuation, and vice versa. Therefore, the surrounding environment of the chestnut area is darker, indicating higher absorption of terahertz signals. Another part of the dehydration process shows that the attenuation of terahertz signals is mainly caused by moisture, and the overall attenuation mainly depends on the overall moisture content in the sample (ignoring the contribution of residual substances).

Using fifty chestnuts as research subjects, they extracted and compared the physical characteristics (weight and volume) of the fruits to obtain the attenuation characteristics of terahertz waves. Each chestnut was weighed, and the thickness of the chestnut was measured with a caliper. The chestnuts were then placed in the terahertz sensing area for optical measurement. First, RGB photos were taken with a reflection camera (EOS 1100D Canon camera), and then terahertz images were obtained with a 100 GHz terahertz imaging system. This process was repeated for all fifty chestnuts, which were then sent to a microbiology laboratory for further detection of fungal presence. The chestnut area in the images obtained from the reflection camera and the thickness of the chestnuts were combined to estimate their volume.

Subsequently, the overall attenuation of the chestnuts was extracted from the terahertz images. Specifically, a dedicated Matlab software program was used for chestnut area detection in RGB images, ellipsoid approximation, ellipsoid volume calculation, overall attenuation extraction, and gradient measurement, among others. In addition, the spatial gradient of attenuation was calculated, and a preliminary strategy for selecting healthy fruits was developed based on it.

The overall attenuation of healthy chestnuts and their weight and volume show a linear relationship. The results show that the overall terahertz attenuation of chestnuts is roughly proportional to the water content in the chestnuts, and previous studies have shown that differences in water content may be directly related to the presence of pathogens. Therefore, infected fruits are affected by changes in water density, while the water density of healthy fruits is almost constant. However, due to the high number of infected fruits, the statistical data of healthy fruits are not representative. The linear relationship can be used to develop a preliminary strategy for screening healthy fruits. In this framework, they considered the ratio of overall attenuation A to weight w (A/w) and the ratio of overall attenuation to volume v (A/v). The distribution of A/w and A/v ratios for all healthy and unhealthy chestnuts was analyzed. The median and mode of the A/w and A/v distributions of healthy chestnuts are higher than those of unhealthy chestnuts. For the A/v distribution, the average value for healthy chestnuts is also higher than that for unhealthy chestnuts.

The A/v distribution of healthy chestnuts is higher than that of unhealthy chestnuts. Since the absorption rate may be roughly proportional to the water content, this result suggests that the overall weight and volume of water content in unhealthy chestnuts are often lower than in healthy chestnuts. To dissolve and assimilate insoluble macromolecules, fungal microorganisms produce and secrete some enzymes from the mycelium, which gradually attack and destroy the macromolecular components of the substrate they colonize (such as through the hydrolysis of polysaccharides). Therefore, in unhealthy chestnuts, the lower water content may be due to the hydrolysis of carbohydrates.

The research team proposed a non-invasive and real-time method for the possibility of individualized detection of pathogens in chestnuts. In traditional methods, the study of chestnuts affected by pathogens requires invasive experimental methods. On the other hand, their method provides a fast and non-destructive way to screen the quality of contaminated fruits. Although the number of healthy chestnuts found in their study is small, these results can serve as a strategy for real-time identification of unhealthy chestnuts, as these chestnuts significantly deviate from the statistical distribution of healthy fruits.

5.8 Impurity Detection in Wheat

Wheat is one of the main food sources for humans and plays a significant role in human life. Impurities such as wheat husks, wheat straw, wheat leaves, wheat grains, weeds, and ladybugs can easily mix with wheat during the harvesting process, reducing the quality of the harvest and affecting the quality of the wheat to some extent. In 2021, researchers from Southwest University and the National Engineering Research Center for Information Technology in Agriculture proposed a method for rapidly and effectively detecting impurities in wheat based on terahertz spectral imaging and convolutional neural networks (CNN) [12]. First, the spectral characteristics of wheat, wheat husks, wheat straw, wheat leaves, wheat grains, weeds, and ladybugs were studied in the frequency range of 0.2–1.6 THz using terahertz spectral imaging technology, and the corresponding frequency domain spectra were obtained using Fourier transformation. Then, the absorption coefficients and refractive indices were calculated. Based on the principle of maximum frequency domain imaging, terahertz pseudocolor imaging was performed on wheat and its impurities, and a new wheat-V2 CNN was designed to extract data and information from spectral imaging features. Finally, the wheat-V2 model was compared with the ResNet-V2_50 and ResNet-V2_101 models under the same conditions. In addition, the experimental results were evaluated using loss function and confusion matrix indicators. The results showed that the designed wheat-V2 model can effectively identify impurities in wheat images, with recognition accuracies of 97.56% and 98.58% for the Top_1 and Top_5 validation sets, respectively. Moreover, the average F1 score for image recognition of various impurities by the designed wheat-V2 model was 97.83%, higher than the traditional models ResNet-V2_50 and ResNet-V2_101.

The detection system mainly consists of two parts—one for data collection and processing, and the other for classification. Measurements were made using a terahertz spectrometer (TEKAK15) produced by Menlo Systems GmbH. After collecting the corresponding data through a computer, the terahertz time-domain waveform was obtained. At the same time, the time-domain waveform can be converted into a frequency-domain waveform through Fourier transformation to complete the scanning of the sample. Finally, CNN was used to process and classify the imaging data to detect wheat and its impurities.

Wheat, wheat husks, wheat straw, wheat leaves, wheat grains, weeds, and ladybugs were selected as samples. All samples were crushed into powder, approximately 200 parts (size 75 μm), and pressed into thin slices mixed with polyethylene powder.

Measurements were made on wheat, wheat husks, wheat straw, wheat leaves, wheat grains, weeds, and ladybugs using a terahertz spectrometer (TEKAK15) produced by Menlo Systems GmbH. After collecting the corresponding data through a computer, the terahertz time-domain waveform was obtained. At the same time, the time-domain waveform can be converted into a frequency-domain waveform through Fourier transformation to complete the scanning of the sample. Terahertz images require a large amount of data, with each pixel corresponding to a time-domain waveform. Any data point's amplitude or phase can be selected from the time-domain spectrum or its Fourier-transformed frequency-domain spectrum for imaging, reconstructing the sample's spatial density distribution, refractive index, and thickness distribution. They created images of the samples by extracting the maximum time-domain spectrum of each sample. Then, the improved CNN wheat-V2 model was used for processing, which includes an input layer, a combination of four convolution and pooling layers, two fully connected layers, and an output layer.

In the time-domain waveform, compared with the reference signal, the maximum peaks of the six impurities all have different degrees of attenuation. Due to the absorption, reflection, and dispersion effects of terahertz waves on the sample surface, the delay time and peak attenuation of different impurities will be different. Because the transmittance of the terahertz pulse signal and the reference signal of the six impurities are different, the refractive index of each impurity will also be different, indicating that each sample has different absorption and scattering for terahertz waves. By performing a Fourier transformation on the terahertz time-domain signal of different impurities, the frequency-domain spectrum can be obtained. It can be seen from the figure that wheat husks, wheat straw, and weeds have characteristic peaks at 0.29 THz and 0.48 THz, and weak characteristic peaks at 0.60 THz, while wheat leaves, ladybugs, and normal wheat grains have obvious characteristic peaks at 0.29 THz. Therefore, these terahertz characteristic peaks can serve as feature fingerprints for wheat impurity detection. Since different impurities have different spectral curves, their absorption coefficients and refractive indices are also different, indicating that different impurities have different physical properties.

After data processing, it was found that the refractive index of wheat grains in the terahertz band is significantly different from that of various impurities. The average refractive index of wheat grains is the highest, indicating that the transmission speed of terahertz waves in wheat grains is slower. The average refractive index of wheat husks is the lowest, indicating that the transmission speed of terahertz waves in wheat husks is faster. Different types of impurities can be distinguished by their corresponding refractive indices. The overall absorption of each sample in the frequency range of 0.2–1.6 THz increases with the increase of frequency. The absorption coefficient of wheat straw is the smallest, and the absorption of normal wheat grains is the strongest, with the highest absorption coefficient, indicating that the composition of normal wheat grains tends to be stable. Wheat husks, weeds, and wheat straw have similar absorption spectra in the terahertz band. Their main components are the same, but the actual measured spectra have significant differences, indicating that the content of cellulose, lignin, hemicellulose, and other main components in various impurity samples is different. Through the qualitative analysis of the spectrum, absorption coefficient, and refractive index parameters, wheat husks, weeds, and wheat straw do not have obvious features in the target terahertz band, while wheat grains, wheat leaves, and ladybugs have obvious characteristics in the same band. For high-density samples, molecules can absorb more terahertz radiation, and the terahertz images of different samples will also be different.

The maximum imaging method includes imaging at each frequency separately, and the distribution and concentration of substances can be determined through the generated terahertz grayscale images. Due to the low recognizability of grayscale images, it is necessary to establish frequency color intervals and use the frequency information within the interval to generate color terahertz images. The distribution and concentration information of different substances can be used to form pseudo-color images with the colors of color pictures. Wheat grains, wheat leaves, weeds, wheat straw, ladybugs, and wheat husks have good imaging effects at 0.29 THz, 0.48 THz, and 0.60 THz frequencies, with clear image textures, visible edge contours, and good recognizability. When the frequency increases to a certain value, the terahertz image becomes more blurred, the noise increases, and the image is almost distorted. In addition, the imaging effect at 1.60 THz frequency is not ideal, the imaging error is large, and the recognizability is low, which is caused by the spatial diffraction limit.

Due to the different imaging effects between different terahertz frequencies, and the large amount of imaging data, it is necessary to filter image features. To avoid the problems of multiple imaging and subsequent image filtering in imaging, CNN was used to post-process the data. The frequency domain images in the frequency range of 0.2–1.6 THz were extracted from the terahertz images and used as the input of CNN after denoising. Each impurity image was normalized to a size of 224×224, a terahertz image dataset of wheat and impurities was created, and each dataset was divided into independent training sets, validation sets, and test sets. To fully train the CNN, improve its generalization performance and robustness, the research team used data augmentation methods to expand sample data and enhance data diversity. They trained the wheat impurity image dataset on ResNet-V2_50, ResNet-V2_101, and Wheat-V2 models respectively. For different impurities, they selected 270 images of each type of impurity as the test set and tested the model separately. The accuracy and recall were combined, and the F1-score was used as the harmonic mean of accuracy and recall. The average F1-scores of Wheat-V2, ResNet-V2_50, and ResNet-V2_101 were 97.83%, 96.59%, and 96.78% respectively, indicating that Wheat-V2 is superior to ResNet-V2_50 and ResNet-V2_101 in terms of F1-score. The recognition rate of the Wheat-V2 network for wheat leaves, wheat grains, and ladybugs is higher than that for wheat straw, weeds, and wheat leaves, indicating that wheat grains, wheat leaves, and ladybugs can be more easily recognized and have more obvious image features.

References

1. Sun Y, Wang J. Progress in the application of terahertz time-domain spectroscopy technology in food, drugs, and environmental fields. Prog Laser Optoelectron. https://kns.cnki.net/kcms/detail/31.1690.TN.20210706.1744.008.html.
2. Liu Y, Xu Z, Hu J, Li M. Research progress on the quality detection of agricultural products by terahertz spectroscopy. Prog Laser Optoelectron. 2021;58(1):63–74.
3. Hu J, Chen R, Xu Z, et al. Research on enhanced detection of benzoic acid additives in liquid food based on terahertz metamaterial devices. Sensors (Basel). 2021;21(9):3238.
4. Liu W, Zhang Y, Li M, et al. Determination of invert syrup adulterated in acacia honey by terahertz spectroscopy with different spectral features. J Sci Food Agric. 2020;100(5):1913–21.
5. Sun X, Liu J. Measurement of plumpness for intact sunflower seed using terahertz transmittance imaging. J Infrared Millim Terahertz Waves. 2020;41:307–21.
6. Otsu N. A threshold selection method from gray-level histograms. IEEE Trans Syst Man Cybern. 1979;9:62–6.
7. Liu L, He G, Wul, et al. Non-destructive detection of tobacco filter capsule by terahertz time domain spectroscopy. 2021 46th International Conference on Infrared, Millimeter and Terahertz Waves (IRMMW-THz). IEEE, p. 1–2.
8. Xia Y, Liu W, Shi Y, et al. Rapid determination of capsaicin concentration in soybean oil by terahertz spectroscopy. J Food Sci. 2022;87:567–75.
9. Yang S, Li C, Mei Y, et al. Discrimination of corn variety using Terahertz spectroscopy combined with chemometrics methods. Spectrochim Acta Part A Mol Biomol Spectrosc. 2021;252: 119475.

10. Sillo, F, Giordano L, Zampieri E, et al. HRM analysis provides insights on the reproduction mode and the population structure of gnomoniopsis castaneae in Europe. Plant Pathol. 2017;66(2):293–303.
11. Girolamo FVD, Pagano M, Tredicucci A, et al. Detection of fungal infections in chestnuts: a terahertz imaging-based approach. Food Control. 2021;123: 107700.
12. Shen Y, Yin Y, Li B, et al. Detection of impurities in wheat using terahertz spectral imaging and convolutional neural networks. Comput Electron Agric. 2021;181: 105931.

Chapter 6
Application of Terahertz Time-Domain Spectroscopy in the Biomedical Field

Biomedical engineering is an interdisciplinary field closely related to bioengineering. It is characterized by the application of engineering and applied science knowledge and techniques to solve scientific problems in biology and medicine, thoroughly studying life systems and their behaviors, and developing related biomedical systems. Spectroscopy is an important branch of biomedical detection, including traditional Raman spectroscopy, Fourier infrared spectroscopy, ultraviolet absorption spectroscopy, etc., often used for high-sensitivity detection of biomarkers in molecular, cellular, tissue and other biological samples.

From the perspective of spectroscopy, due to the energy levels of many interactions between biological macromolecules being in the terahertz band, they exhibit strong absorption and resonance in the terahertz band, such as the stretching vibration, deformation vibration, and bending vibration of macromolecules. Terahertz spectroscopy can detect information about interactions between macromolecules that other frequency spectra cannot detect. As shown in Fig. 6.1, terahertz spectroscopic imaging currently has many advanced applications in biomolecular, cellular, bacterial, and tissue spectroscopic imaging. Studies have shown that various weak interactions within or between various biological macromolecules in cells, including RNA, DNA, amino acids, proteins, and polypeptides (mainly covering hydrogen bonds and van der Waals forces), and vibrations or rotations between skeletons, are precisely in the terahertz band. Terahertz time-domain spectroscopy can be used to characterize various biological macromolecules.

The wavelength of terahertz waves is long, and the photon energy is far less than the energy of X-rays, so it will not ionize biological macromolecules, cells, and tissues. Compared with other medical imaging technologies, terahertz wave spectroscopic imaging technology has more unique and applicable physical characteristics. In addition, water strongly absorbs terahertz, and terahertz waves are extremely sensitive to biological water content. This feature can be an important means of studying the distribution of water in organisms. For example, due to abnormal angiogenesis and metabolic circulation around tumor cells, the water content in their cells will be

T. Chen et al., *Terahertz Time Domain Spectroscopy*,
https://doi.org/10.1007/978-981-95-2605-5_6

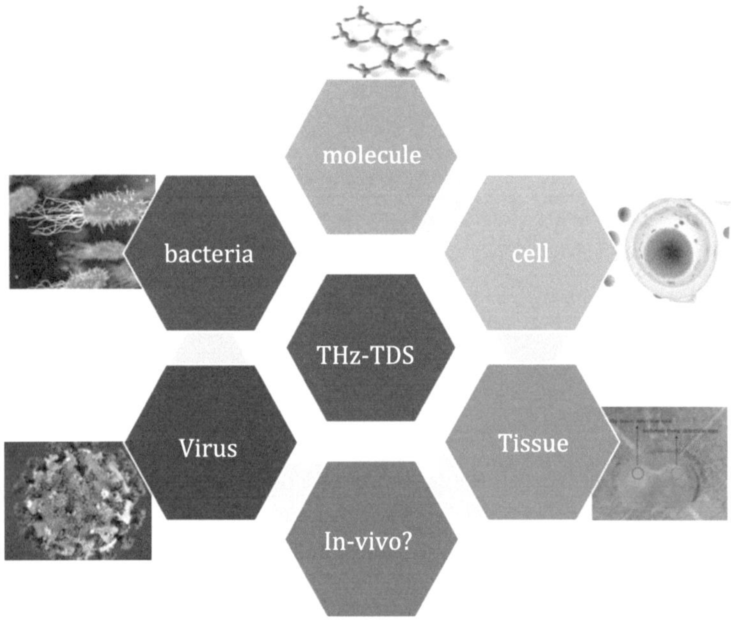

Fig. 6.1 Terahertz biomedical application

significantly higher than that of normal cells. The detection of cell water content by terahertz time-domain spectroscopy has become an important basis for distinguishing tumor cells from normal cells.

Internationally, the European Union established the "THz-Bridge" international joint project in 2000, and Europe has made biomedical applications a top research direction for ten consecutive years. In China, the 488th Xiangshan Science Conference themed "Scientific Issues and Frontier Technologies of terahertz Waves in Biomedical Applications" was held in April 2014. Research on terahertz biomedicine has been fully rolled out and is flourishing. Terahertz time-domain spectroscopy provides a new measurement and analysis method for biomedical research. In this chapter, we focus on the application of terahertz time-domain spectroscopy in biomolecules, cells, bacteria, tissues, etc. Terahertz virus detection has just begun to be reported, and the latest results have attracted widespread attention from international research. Due to the influence of the water content of the organism, it is difficult for terahertz time-domain technology to be directly applied to in vivo detection, which will not be elaborated in this chapter.

6.1 Detection of Biological Macromolecules

In the filed of biomedical microcosm, the terahertz technology can be used to reveal the interaction rules between large biological molecules such as ribonucleic acid (RNA), amino acids, proteins, etc., reflecting the physical characteristics of the interactions and activities between these large molecules. This provides a revolutionary scientific method for explaining various life phenomena.

6.1.1 Detection of MicroRNA

Due to the weak interactions, rotations, and vibrational energy levels between various large biological molecules such as nucleic acids, which are just in the terahertz band, terahertz waves are very sensitive to the structural changes of nucleic acid molecules. Therefore, terahertz time-domain spectroscopy can effectively detect structural changes and biological reactions between nucleic acid molecules. Specifically, terahertz spectroscopy can detect the functional status of nucleic acids, such as the interaction between nucleic acid molecules and drugs detected by terahertz time-domain spectroscopy, and the dynamics of nucleic acids. As the most important carrier in organisms, ribonucleic acid is often an important marker of gene expression, cell cycle, and organism development. As a typical example, we first introduce the detection of microRNA by terahertz time-domain spectroscopy [1].

MicroRNA, also known as miRNA, is a short, highly conserved non-coding ribonucleic acid molecule that naturally exists in the genomes of plants and animals. Their length is 17 to 27 nucleotides. MicroRNA is transcribed from deoxyribonucleic acid (DNA), but it cannot be further translated into protein-coding RNA, so it is a non-coding RNA. MicroRNA regulates post-transcriptional gene expression by binding to target messenger RNA (mRNA). Studies have shown that thousands of human protein-coding genes are regulated by microRNA, and microRNA is a major regulator and important marker of many important biological processes. MicroRNA plays an important role in gene regulation, development differentiation, and apoptosis necrosis in tumor cells. Abnormal expression of various microRNA molecules has been detected in the body fluids of various tumor patients, such as in the serum of metastatic breast cancer patients and the plasma of early liver cancer patients. Therefore, microRNA molecules are widely considered as potential tumor markers, which can be used in the diagnosis and treatment process of various tumors and post-treatment observation.

Due to the low concentration and short sequence homology in microRNA samples, the precise quantification of microRNA still poses a great challenge. The content of microRNA in human blood samples is very low, such as the concentration in plasma is about 0.6×10^{-30} pg/ml. The sequence length of microRNA is short, and the difficulty of amplification is greater than that of general RNA. This short-chain RNA, microRNA, is easily lost during the sample extraction process. Many microRNA molecules have high sequence homology, which reduces the specificity of detection. Establishing a new method for rapid, simple, and highly sensitive quantification of microRNA is of great significance for the diagnosis and prognosis of tumors. By combining terahertz spectroscopy time-domain and nucleic acid amplification methods, microRNA can be quantitatively detected.

Strand displacement amplification (SDA) is an enzyme-catalyzed DNA isothermal amplification method developed in recent years. Strand displacement amplification can specifically amplify the target microRNA molecule and is currently widely used in the detection research of various microRNA molecules. Its basic principle is to carry chemically modified restriction endonuclease recognition sequences at both ends of the target DNA. The endonuclease opens a gap at its recognition site on the DNA strand, and the DNA polymerase extends the gap end and replaces the next DNA strand. The replaced DNA single strand can bind to the primer and be extended into a double strand by DNA polymerase. This process is repeated continuously, resulting in efficient amplification of the target sequence.

Taking breast cancer-specific microRNA-21 as an example, its amplification process is as follows: first, miRNA-21 is hybridized with template DNA, and the DNA strand is extended using Klenow fragment polymerase. After forming a complete double-stranded DNA, the extended DNA strand is cut with a nicking endonuclease, and the Klenow fragment then initiates a new round of replication, with the first replicated DNA strand being displaced and released. After repeated many times, a large amount of trigger DNA is produced. Then, the thiol probe-modified nanoparticles are used to form nano probes, which are hybridized with the complementary fragment of the trigger DNA chain and the biotin probe to form a biotinylated nanoparticle-trigger DNA complex. The biotinylated complex is captured by binding to streptavidin-coated magnetic beads. After magnetic separation and denaturation, the pure nanoparticle-trigger DNA complex is separated and finally dropped onto the metamaterial surface for terahertz spectroscopy measurement. The amplification and measurement process of microRNA is shown in Fig. 6.2.

Nanoparticles, as signal enhancement components, have a significant impact on the sensitivity of terahertz metamaterial sensors due to their diameter and material. To explore this impact, nanoparticles of gold and silver were selected, with three different diameters for each material, totaling six types of nanoparticles. Their response to the terahertz time-domain spectroscopy signal after binding with the chain displacement amplification product of microRNA-21 on the metamaterial was analyzed. The normalized terahertz transmission spectra of the composite samples of AuNPs gold nanoparticles and AgNPs silver nanoparticles with diameters of 20 nm, 50 nm, and 100 nm bound to trigger DNA are shown in Fig. 6.3. The terahertz spectroscopy results show that, under the same diameter, AuNPs gold nanoparticles can produce a larger metamaterial resonance peak frequency shift (Δf) than AgNPs silver nanoparticles, corresponding to a higher detection sensitivity. Under the same material, 50 nm diameter nanoparticles can produce a larger resonance peak frequency shift Δf than other diameters. It can be seen that 50 nm AgNPs gold nanoparticles are the best conditions for signal enhancement, producing the largest resonance peak offset.

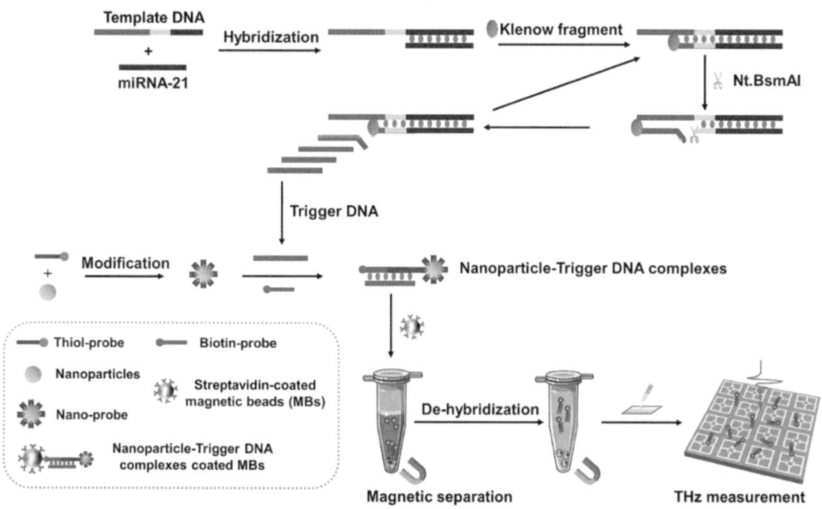

Fig. 6.2 miRNA amplification and measurement process. Reprinted with permission from Elsevier from Ref. [1], ©2021. Permission conveyed via RightsLink, licence no. 6078700391734, 30 July 2025

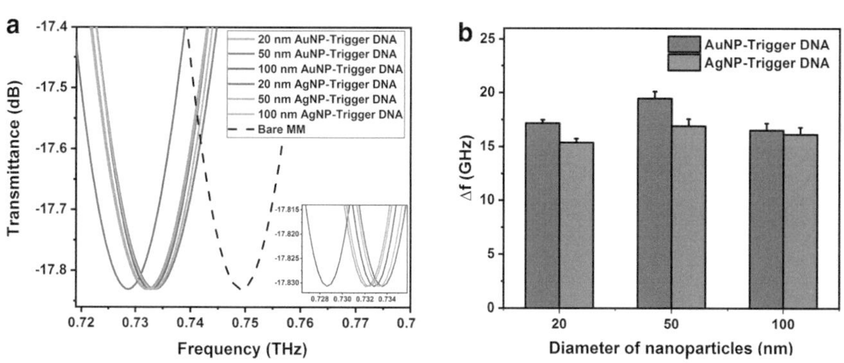

Fig. 6.3 **A** Terahertz time-domain spectral response of blank metamaterials and DNA samples with different diameters of AuNPs gold nanoparticles and AgNPs silver nanoparticles **B** Frequency shift of DNA samples combined with different diameters of AuNPs and AgNPs. Reprinted with permission from Elsevier from Ref. [1], ©2021. Permission conveyed via RightsLink, licence no. 6078700391734, 30 July 2025

Under the best signal enhancement conditions, 50 nm diameter AgNPs gold nanoparticles bind with DNA triggered by different concentrations of microRNA-21, used to measure the detection sensitivity of terahertz metamaterial biosensors. The experimental results show that to further measure the relationship between the concentration of miRNA-21 and the response of terahertz spectroscopy, the concentration range of microRNA-21 linearly increases from 10 aM to 100 pM. As the

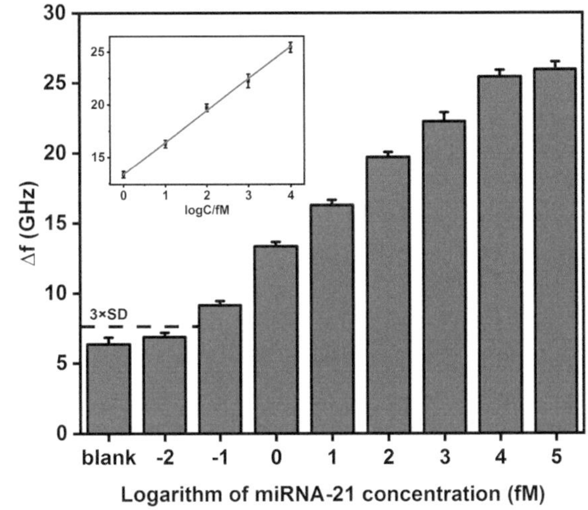

Fig. 6.4 Linear change in terahertz resonance peak shift with different concentrations of microRNA-21 using metamaterial sensors. Reprinted with permission from Elsevier from Ref. [1], ©2021. Permission conveyed via RightsLink, licence no. 6078700391734, 30 July 2025

concentration of microRNA-21 increases, the terahertz spectroscopy resonance peak offset Δf also continues to increase. The logarithm of the concentration of microRNA-21 and the terahertz spectroscopy resonance peak offset Δf are linearly fitted, and the results are shown in Fig. 6.4. When the concentration of microRNA-21 is in the range of 1 fM-10 pM, there can be a good linear relationship between its logarithm and Δf. The linear regression equation is $\Delta f = 3.04 \log c + 13.37$ ($R^2 = 0.9985$). According to the method of three times the standard deviation plus blank, the limit of detection of microRNA-21 can be calculated to be 14.54 aM. The research results show that the terahertz metamaterial sensor combined with terahertz time-domain spectroscopy measurement technology has a high detection sensitivity to microRNA, and has important application prospects in nucleic acid analysis and tumor diagnosis.

6.1.2 Amino Acid Detection

Amino acids are important organic compounds in biology, composed of amino ($-NH_2$) and carboxyl (–COOH) functional groups, as well as side chains attached to each amino acid. Amino acids are the basic units that make up proteins, giving proteins their molecular structure and biochemical activity. Analyzing the characteristics of amino acids is of great significance for proteins and biomolecules. Due to space limitations, this section will focus on the terahertz time-domain spectroscopy response of phenylalanine and levodopa.

Due to the different spatial arrangement of carbon atoms inside amino acids, amino acids are divided into different chiralities: D-type, L-type, and DL-type. The amino acids that make up natural proteins are all L-type. Among them, D-type and L-type amino acids are enantiomers, featuring the same structure and being chiral mirror

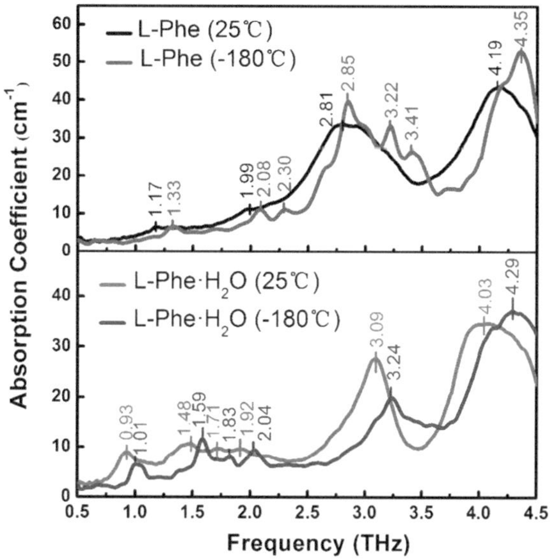

Fig. 6.5 Terahertz spectrum of L-phenylalanine and its hydrate. Reprinted with permission from Elsevier from Ref. [2], ©2021. Permission conveyed via RightsLink, licence no. 6078680450049, 30 July 2025

images, and DL-type amino acids belong to racemic forms. They interact differently and have completely different biological functions and meanings in biology [24]. Using the sensitivity of terahertz wave to hydrogen bonds, the molecular spectra of amino acids and their hydrates can be studied. Figure 6.5 shows the terahertz spectra of L-phenylalanine and its hydrates. At 25 °C, L-phenylalanine has strong absorption peaks at 2.81 THz and 4.19 THz. When cooled to −180 °C (93 K), the absorption peaks show a blue shift. At 25 °C, the terahertz absorption peaks of L-phenylalanine hydrates are at 3.09 THz and 4.03 THz. At −180 °C (93 K), L-phenylalanine hydrates have strong absorption peaks at 3.24 THz and 4.29 THz [2]. Density functional theory calculations show that the terahertz spectroscopy absorption characteristics of phenylalanine monohydrate mainly come from the hydrogen bond interaction between phenylalanine and water molecules.

In order to further analyze the relationship between the terahertz absorption spectrum and vibration mode of L-phenylalanine, the experimental measurement and simulation calculation spectra of L-phenylalanine at 25 °C and −180 °C are compared, as shown in Fig. 6.6a. The vibration mode calculated by simulation shows that the absorption peak at 1.91 THz is produced by the translation of L-phenylalanine, while the absorption peak at 3.15 THz is produced by the rotation of L-phenylalanine, as shown in Fig. 6.6b.

L-DOPA, also known as β-3,4-dihydroxyphenyl-α-alanine, is a precursor derivative of important neurotransmitters such as adrenergic and dopamine, and is also a metabolic product of tyrosine in the body. Within the central nervous system in humans, L-DOPA is converted into dopamine by dopa decarboxylase, hence its physiological function is closely related to neural transmission in the nervous system. L-DOPA is also a key drug in the treatment of Parkinson's disease and has

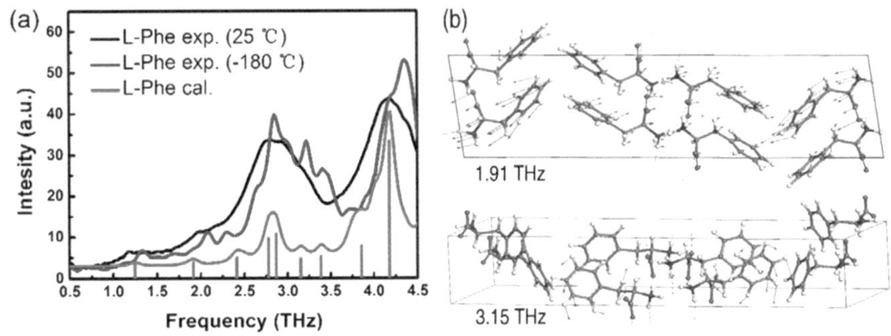

Fig. 6.6 **a** Simulated and experimentally measured terahertz spectrum of L-phenylalanine. **b** Vibrational modes of L-phenylalanine at 1.91 THz and 3.15 THz. Reprinted with permission from Elsevier from Ref. [1], ©2021. Permission conveyed via RightsLink, licence no. 6078700391734, 30 July 2025

significant clinical applications. The L-DOPA molecule consists of a rigid phenyl ring and a flexible side chain, where the two adjacent phenol hydroxyl groups on the phenyl ring and the amino and carboxyl groups on the side chain are functional groups. The hydrogen bond network in these functional groups maintains the molecular structure. The hydrogen bonds in L-DOPA play a crucial role in its biological activity, affecting the conformation and biological function of L-DOPA. Terahertz time-domain spectroscopy can characterize the hydrogen bond interactions between L-DOPA molecules, detect the fingerprint spectra of biomolecules in the terahertz band, and achieve label-free detection of L-DOPA molecules. Using terahertz time-domain spectroscopy and Fourier transform infrared spectroscopy, the terahertz absorption spectra of 0.5–14 THz can be characterized as shown in Fig. 6.7, showing that L-DOPA has a series of terahertz absorption peaks [3].

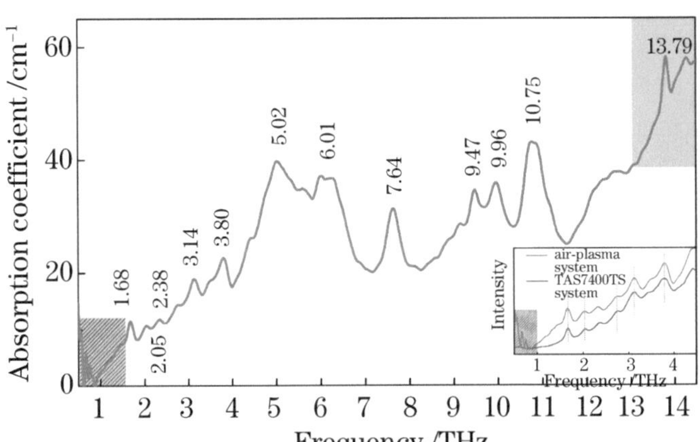

Fig. 6.7 Terahertz absorption spectrum of levodopa molecules. Reprinted from Ref. [4], ©2010, under CC-BY 4.0

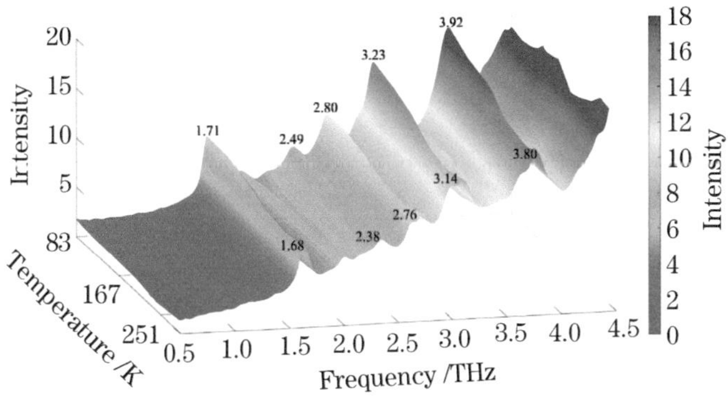

Fig. 6.8 Response of Levodopa terahertz spectrum to measurement temperature. Reprinted from Ref. [3], ©2019, under CC-BY 4.0

Changes in measurement temperature can cause changes in the resonance absorption peaks. The measurement temperature of L-DOPA varies within the temperature range of 83–273 K. The response spectrum of the terahertz absorption coefficient to changes in measurement temperature is shown in Fig. 6.8 [5]. From the results, it can be seen that the terahertz spectral absorption peaks at room temperature are located at 1.68 THz, 2.38 THz, 2.76 THz, 3.14 THz, and 3.80 THz. As the temperature decreases, the absorption peak positions shift to varying degrees of blue shift. When the temperature drops to 83 K, the absorption peaks shift to Δf of 1.71 THz, 2.49 THz, 2.80 THz, 3.23 THz, and 3.92 THz, and each peak frequency shift varies. Multiple mechanisms contribute to the movement of the terahertz absorption spectrum of L-DOPA molecules. It is generally believed that when the temperature decreases, the blue shift of the peak position is mainly related to the non-harmonic vibration of the hydrogen bond and the contraction of the lattice volume.

6.1.3 Protein Molecule Detection

Proteins are essential components of cells and tissues in organisms on Earth, participating in every process of all important life activities. Enzymes are the most common proteins, catalyzing biochemical reactions and being crucial for the metabolism of organisms. In addition to enzymes, many structural or mechanical proteins, such as actin and myosin in muscles, and tubulin in the cytoskeleton, participate in forming the intracellular support network to maintain cell shape, which is vital for organisms. Other proteins are involved in cell signal transduction, immune response, cell adhesion, and cell cycle regulation. Proteins are the material basis of life and large organic molecules, constituting the basic organic matter of cells and the main bearers of life activities. Proteins are mostly large biomolecules. Terahertz time-domain spectroscopy can characterize the spectral characteristics of proteins, thereby providing new insights for biological effects.

Protein aggregation is a biological phenomenon where abnormally folded proteins accumulate and aggregate inside or outside cells. These protein aggregates are often associated with diseases. Protein aggregation is a biological phenomenon where abnormally folded proteins accumulate and aggregate inside or outside cells, and protein aggregation is a mechanism of great importance for regulating cell proliferation and survival. Proteins usually undergo conformational changes and accumulate through the aggregation process, eventually forming irreversible amyloid fiber structures. Protein aggregates are closely related to various diseases, especially neurodegenerative diseases related to amyloidosis, including Parkinson's syndrome, Alzheimer's disease, and Huntington's chorea. The intermolecular forces between protein molecules drive the aggregation of multiple protein molecules, while also maintaining the stability of the structure and morphology of the protein after aggregation, playing an important role in the protein aggregation process.

Insulin is a protein that can promote liver cells and muscle cells to convert glucose into glycogen. Insulin is used clinically to treat hyperglycemia in diabetic patients. Insulin easily aggregates in vitro, forming different aggregates under different conditions, which can be artificially controlled. The insulin monomer molecule is a three-dimensional structure formed by folding a long chain of polypeptides, with an irregular shape and uneven surface. In their natural state, insulin molecules do not exist as single molecules. At physiological pH values (7.35–7.45) Under certain conditions, two insulin monomer molecules can easily aggregate to form a dimer, and two dimers can further aggregate to form a tetramer or hexamer. Terahertz spectroscopy was performed on solid-state insulin monomers and insulin aggregates, with the results shown in Fig. 6.9. The terahertz absorption of insulin aggregates is higher than that of insulin monomers, and the refractive index of insulin aggregates in the terahertz band is also higher than that of insulin monomers, indicating that terahertz spectroscopy can effectively distinguish between insulin monomers and insulin aggregates [4].

Insulin typically exists in a liquid environment, so we further analyzed the terahertz spectroscopic absorption characteristics of insulin solutions. Terahertz spectroscopy was performed on insulin monomer solutions and dimer solutions. The insulin concentration varied uniformly within the range of 0–29 mg/ml, with a concentration gradient of 2.9 mg/ml. Figure 6.10 shows the terahertz absorption coefficient curves for insulin monomer solutions (a) and insulin dimer solutions (b) at different concentrations [6]. At lower concentrations, the absorption coefficient increases with increasing concentration, but beyond a certain point, further increases in concentration actually lead to a decrease in the absorption coefficient. This is generally believed to be due to the interaction between water molecules in the hydration layer of insulin and the protein, which leads to a preferential orientation of water molecules, resulting in an increase in density. The terahertz absorption of the insulin hydration layer is greater than that of free water.

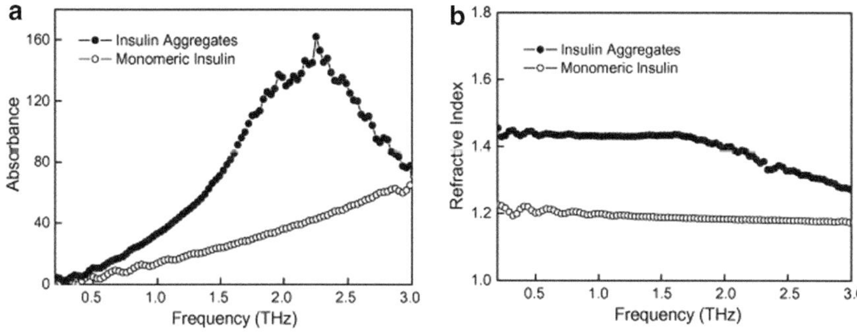

Fig. 6.9 Terahertz absorption coefficient spectrum (**a**) and refractive index spectrum (**b**) of insulin monomer and insulin polymer. Reprinted with permission from Elsevier from Ref. [4], ©2010. Permission conveyed via RightsLink, licence no. 6079090725135, 30 July 2025

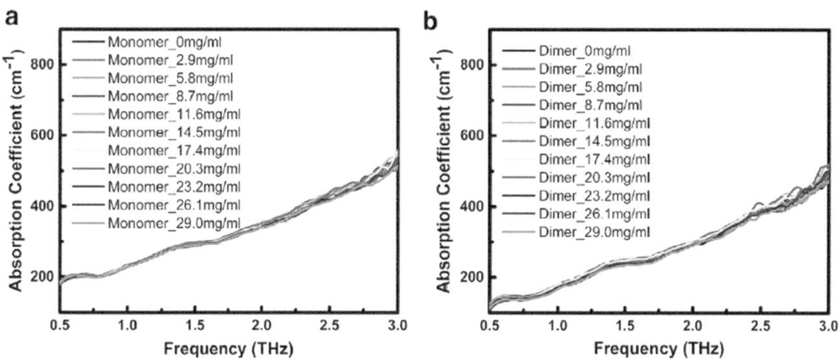

Fig. 6.10 Terahertz absorption spectrum of insulin monomer solution (**a**) and Insulin Dimer Solution (**b**). Reprinted from Ref. [6], ©2019, under CC-BY 4.0

To assess the stability of terahertz spectroscopy, the inflection point concentrations in the sub-frequency band were chosen to compare. The difference between the terahertz absorption coefficient of the insulin solution and the solvent terahertz absorption coefficient is defined as the change in the terahertz absorption coefficient of the insulin solution, $\Delta\alpha$. Considering the poor signal-to-noise ratio of terahertz spectroscopy at high frequencies, we chose to investigate the trends in the absorption coefficient difference $\Delta\alpha$ and solution concentration within the 0.8–1.2 THz and 1.3–1.7 THz sub-frequency bands, as shown in Fig. 6.11.

By averaging the inflection point concentrations within the sub-frequency band, the difference between the inflection point concentrations of insulin monomers and insulin dimers can be analyzed. The maximum absorption coefficient of insulin monomers corresponds to a concentration of about 11.19 mg/ml, with a corresponding insulin monomer particle concentration of 1.93 mM and a diameter of 27.9 Å; the inflection point concentration of insulin dimers is 14.47 mg/ml, with a

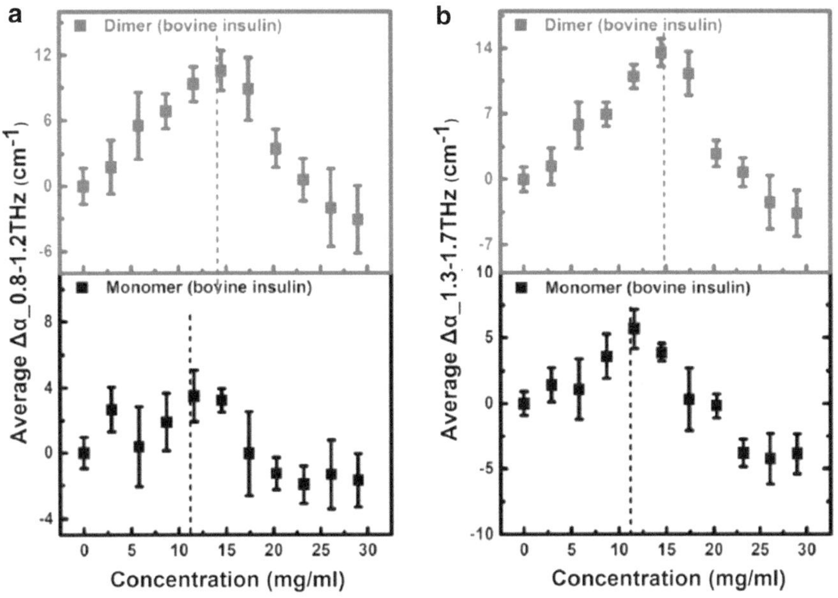

Fig. 6.11 Variation of terahertz absorption coefficient in sub-bands for insulin (Black) and insulin dimer (Red) solutions. Reprinted from Ref. [6], ©2019, under CC-BY 4.0

corresponding insulin particle concentration of 1.25 mM and a dimer insulin molecule diameter of 34.5 Å. The inflection point concentration corresponds to the situation where the hydration layers of adjacent biomolecules just start to overlap. Starting from the inflection point concentration and combining with the random distribution model, information about the thickness of the hydration layer can be quantitatively obtained. The thickness is determined to be 15.4 Å for insulin monomers and 17.5 Å for insulin dimers. Terahertz spectroscopy also has a high sensitivity to structural differences in insulin solution samples.

6.2 Pathogen Detection

Bacteria are one of the main groups of organisms and are the most numerous of all organisms. They have a significant impact on human activities. On the one hand, humans utilize bacteria, for example in the production of cheese, yogurt, and fermented beverages, the manufacture of certain antibiotics, and wastewater treatment. On the other hand, bacteria are the pathogens of many diseases and can spread diseases among healthy individuals through contact, the digestive tract, the respiratory tract, insect bites, and other means. They are highly infectious and pose a great threat to society. Infectious diseases caused by the invasion of pathogenic microorganisms into the human body are still a major threat to human health.

Currently, the clinical detection of pathogens mostly uses traditional microbial identification methods to diagnose infectious diseases. These traditional microbial identification methods, which are based on phenotypic and biochemical tests after microbial culture, are time-consuming, and it usually takes several days to a week to identify the species of the infecting pathogen. If the pathogenic microorganism cannot be diagnosed quickly and accurately, patients with infectious diseases can only receive empirical antimicrobial treatment, rather than targeted antibiotic treatment. The ability to quickly and accurately identify different types of bacteria has important clinical significance.

Microbial contamination has become a prominent food safety issue both domestically and internationally. Diseases caused by microbial contamination pose a serious threat to human health and life, with bacterial food poisoning being the most reported and affecting the most people. Therefore, rapid and accurate monitoring of foodborne pathogens is of great importance. The Guangdong Academy of Agricultural Sciences and Zhejiang University have jointly developed a terahertz microfluidic biosensor based on a metasurface-graphene heterostructure, which enables rapid, accurate, and label-free detection of Escherichia coli DNA [7].

Escherichia coli is the most common and abundant bacteria in the intestines of humans and animals, mainly parasitizing in the large intestine. Generally, E. coli is not pathogenic and can synthesize vitamins B and K, and has bactericidal properties, which are beneficial to the human body. However, certain strains of E. coli can be toxic and cause food poisoning, usually due to the consumption of contaminated meat. Some serotype strains are highly pathogenic, causing diarrhea, collectively referred to as pathogenic E. coli. E. coli needs to be detected quickly and sensitively. Metasurface-assisted terahertz real-time and label-free biosensors have attracted widespread attention, but high-sensitivity specific detection of highly absorbent liquid samples still poses a challenge in the terahertz band. To solve these problems, the research team combined a metal hole array structure with graphene and integrated it into a terahertz microfluidic sensor, establishing a sensing method that converts the concentration of E. coli biological signals into terahertz spectral signal gain. The sensing detection schematic is shown in Fig. 6.12. The terahertz graphene-metasurface microfluidic sensor can effectively reduce the amount of sample solution to be tested, enhance the interaction between biomolecules and terahertz waves, and thus improve detection sensitivity.

In the experiment, specific aptamers were modified on the graphene metasurface, which could recognize the DNA sequence of the foodborne E. coli pathogen O157:H7. The microfluidic sensor with a metal hole array and graphene was used for terahertz spectroscopy detection of different concentrations of E. coli (O157:H7) DNA sequences. The detection results are shown in Fig. 6.13. The lowest transmittance result of the terahertz spectrum can detect 100 nMol of short DNA sequences. Based on the metal hole array and graphene microfluidic biosensor combined with terahertz spectroscopy, the DNA sequence of E. coli can be detected quickly and without labels. This achievement provides a new method for rapid screening and

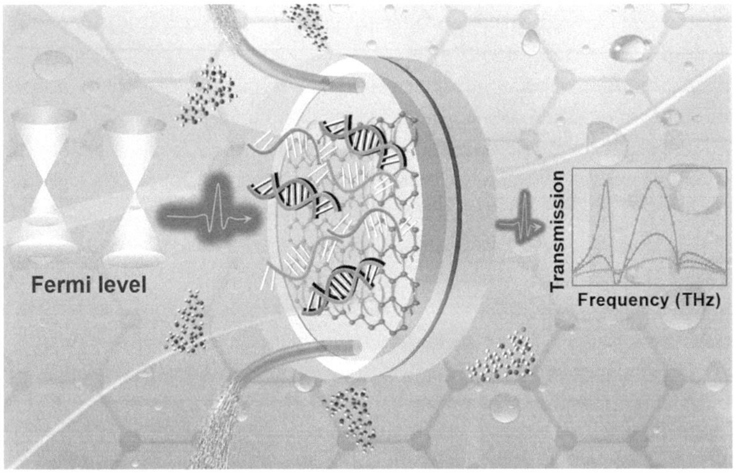

Fig. 6.12 Principle of microfluidic terahertz sensing detection based on metal hole array and graphene. Reprinted with permission from Elsevier from Ref. [7], ©2021. Permission conveyed via RightsLink, licence no. 6079110546492, 30 July 2025

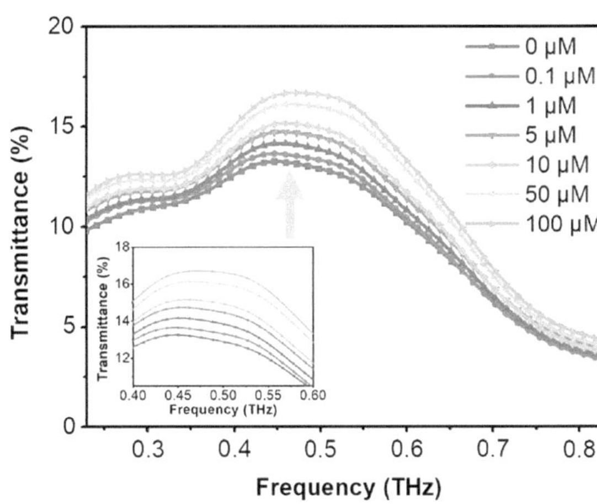

Fig. 6.13 Biosensing of different concentrations of Escherichia coli O157:H7 NDA sequence using metal hole array and graphene microfluidic sensor. Reprinted with permission from Elsevier from Ref. [7], ©2021. Permission conveyed via RightsLink, licence no. 6079110546492, 30 July 2025

detection of foodborne pathogens, and also provides a new approach to solve the common problems of low sensitivity and severe interference from polar matrices in the process of terahertz sensing applications.

6.3 Cell Detection

Cells are the basic structural and functional units of organisms. Apart from viruses, all other organisms are composed of cells, and different cells have different structures and perform different functions. In terms of cell detection technology, it can be divided into labeled and label-free detection based on whether markers are used. Label-free detection technologies mainly include X-rays, CT, MRI, ultrasound, etc., while labeled detection technologies mainly include molecular tomography, multi-photon imaging, and flow cytometry, etc. Currently, labeled detection technology is still the mainstream method in the field of tumor cell detection. This type of technology uses various chemical and biological labeling techniques, such as fluorescence and radionuclide molecules, to reveal various pathological and physiological characteristics of tumor cells with high sensitivity, demonstrating its important value in the field of biomedical applications. Although labeled detection technology has high sensitivity, its sample preparation is complex, and it has certain damage to ex vivo or in vivo living cells. For instance, cytotoxicity, which typically involves the detection of the collective behavior of biomacromolecules or cells, cannot measure the impact of multiple or various biomacromolecular structures and functional changes on the pathophysiological characteristics of living cells at the same time. Compared to labeled detection techniques, label-free detection techniques are more convenient in sample preparation and detection process as they do not require the participation of various markers, and they significantly reduce the damage of markers to cells. However, the drawback of existing label-free detection techniques is that they usually struggle to obtain abundant cell information, and their detection performance needs to be further improved. The use of terahertz technology to measure various types of cells has become one of the research hotspots in the world today. The liquid environment in which cells live and the large amount of liquid water contained within cells have a significant impact on cell metabolism.

6.3.1 Whole Blood Cell Detection

Reid and Jeong, among others, conducted research on the use of THz spectroscopy to detect blood cells during the same period. Reid and his team from the UK were the first to report the spectroscopic measurement results of human whole blood samples [8]. They measured the transmission spectra of whole blood, plasma, blood cells, and thrombi, and determined the absorption coefficients and refractive indices of each component. The absorption of pure water is higher than that of whole blood or its components, and the overall absorption coefficient decreases as the water content in the blood components decreases. However, the water content of blood clots is the same as that of whole blood. The Ryou effect and thrombus formation affect the absorption coefficient. The formation of a thrombus may be a structure of protein/

fibrin covalent binding, which encapsulates a large amount of water, reducing the ability of the sample to absorb terahertz waves.

The refractive index of water is higher than that of whole blood and blood components. Components with high water content have refractive indices very similar to pure water, although this is the case for all whole blood samples, the results for plasma samples are different. The refractive index of plasma is consistent with water and whole blood in the 0.1–0.5 THz low-frequency range, but there are differences in the high-frequency range. The refractive indices of blood clots and blood cells are always 0.05 lower than that of pure water. The absorption and refractive index spectra of whole blood and its thrombi, as well as the Debye coefficients, are significantly different. This indicates that terahertz time-domain spectroscopy measurements are very sensitive to the binding mechanism of whole blood. Due to the high absorption of whole blood and plasma, it is necessary to minimize the presence of blood during terahertz time-domain spectroscopy cell detection to avoid interfering with the diagnostic ability and accuracy of tissue differentiation.

Heong et al. from Yonsei University further demonstrated that terahertz time-domain spectroscopy technology can be used for quantitative analysis of blood [9]. They characterized the optical properties of blood and its components, including water, plasma, and red blood cells, in the terahertz frequency range. The ratios of the optical constants of these components are the same as their volume fractions. In physiological saline and whole blood samples, there is an excellent linear relationship between terahertz absorption and red blood cell concentration. As shown in Fig. 6.14, as the concentration of red blood cells decreases, terahertz absorption increases. At frequencies of 0.5 THz, 1.0 THz, and 1.5 THz, the absorption coefficient decreases linearly with the increase in red blood cell concentration, with R^2 values of 1.00, 0.98, and 0.97, respectively. The red blood cell volume fraction measured by this method is consistent with the traditional blood cell counter.

The above results lay the theoretical foundation for the quantitative detection of cells using terahertz time-domain spectroscopy. Although terahertz spectroscopy can be used to detect the status and quantity of cells, it is difficult to avoid the interference of the absorption of the detection signal by the culture medium layer around the cell sample. The main component of the culture medium around living cells is water molecules, a liquid that is a necessary environment for most cells to survive. Since water molecules are polar, they strongly absorb terahertz waves. Therefore, the culture medium layer around the cell sample will cause attenuation of the terahertz detection signal. In addition, the main components of both the cell sample and the surrounding culture medium are water molecules, meaning that the signals from the cells and the surrounding culture medium are both dominated by water molecules. When using terahertz spectroscopy to detect cells in the culture medium, the terahertz signals from the cells and the culture medium are very close, so the terahertz signal from the cells will be masked by the background signal caused by the culture medium. The problem of recognizing substances in solution becomes very difficult due to the extreme sensitivity of water molecules to terahertz waves, a problem known as the "water sensitivity problem" in the terahertz detection process. Since most cell samples are in a water phase environment and there is also interference

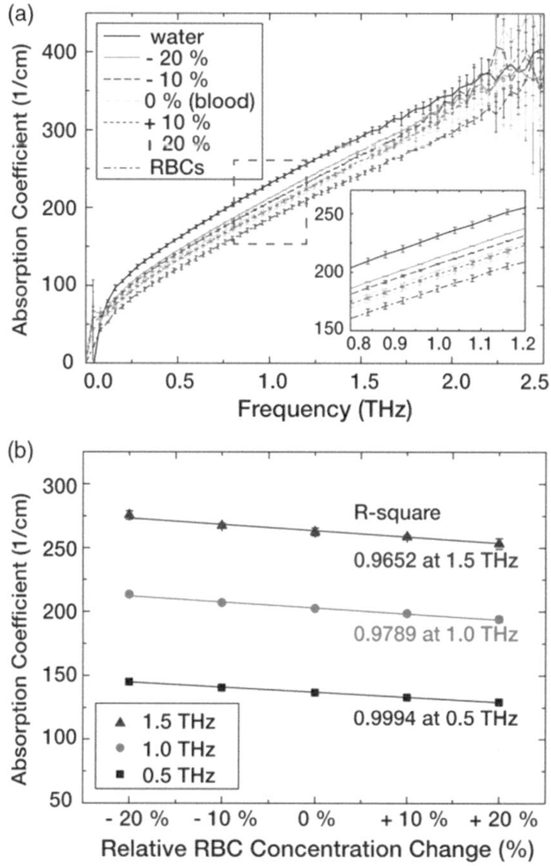

Fig. 6.14 Terahertz absorption spectrum of different concentrations of red blood cells. Reprinted from Ref. [9], ©2013, under CC-BY 4.0

from water vapor in the detection environment, water sensitivity is difficult to avoid in terahertz wave detection of cell samples. Considering that most biological samples from the human body are rich in water or must be in a liquid environment, the detection of living cells by terahertz waves must overcome the interference and obliteration of the detection signal by water sensitivity.

Under normal circumstances, the measurement of terahertz time-domain spectroscopy is completed in about five to ten minutes [10], while the typical time scale of evolution within living cells is in milliseconds or even picoseconds. Therefore, traditional terahertz time-domain spectroscopy detection cannot be easily used for biological and medical examinations. The terahertz single-measurement system of wavefront tilt technology has achieved transient detection of the terahertz pulse time-domain waveform of living HeLa cells within the picosecond time range [11]. $LiNbO_3$ crystals are used to generate terahertz waves, enhancing the radiation power of the terahertz source, thus enabling the detection of water-containing living cells using transmitted terahertz waves. A CCD camera is used as a detector, which can obtain effective measurement data in a single measurement, thus avoiding the time extension of mechanical scanning in traditional terahertz time-domain spectroscopy. The terahertz pulse single-detection technology is based on the principle of photoelectric sampling. After the detection light is diffracted by the blazed grating, the front of the pulse tilts and is focused by the lens after being polarized. Then the detection beam and the terahertz wave are incident on the ZnTe crystal in the vertical direction, and its polarization state changes with the intensity modulation of the terahertz field, and is detected by the half-wave plate and the polarizer. Finally, the signal is amplified by the lens, and the image is recorded by the CCD camera. In the whole process, there is no need to adjust the mechanical delay and perform multiple averages, thus achieving high-frequency, high-resolution terahertz pulse transient detection. Based on this system, they achieved the detection of living cervical cancer cells cultured in vitro and obtained the terahertz characteristic absorption spectrum of living HeLa cells.

6.3.2 Cell Detection Based on Attenuated Total Reflection Terahertz Time-Domain Spectroscopy

Due to the high absorption of water in the terahertz frequency range, both transmission and reflection terahertz spectroscopy measurement techniques cannot be efficiently applied to characterize samples dissolved in polar liquids. Using attenuated total reflection (ATR) measurements can maintain cell activity for a longer time, avoid cell water interference to some extent, and achieve higher accuracy. In the ATR Terahertz Time-Domain Spectroscopy system, the terahertz pulse undergoes total internal reflection on the surface of the ATR prism, generating an evanescent wave that can penetrate a single layer of living cells. On the other hand, ATR Terahertz Time-Domain Spectroscopy has higher sensitivity when measuring high absorption media than transmission or reflection terahertz spectroscopy and mid-infrared ATR spectroscopy [12]. ATR Terahertz Time-Domain Spectroscopy can effectively analyze the terahertz characteristics of living cells cultured in a water-rich medium. The optical path of ATR Terahertz Time-Domain Spectroscopy measurement is shown in Fig. 6.15 [13]. The incident terahertz pulse propagates through the silicon prism and undergoes total reflection at the prism-sample interface at a certain incident angle. Part of the terahertz wave extends into the sample and forms an evanescent wave, the amplitude of which decays exponentially with the depth of incidence. The ATR spectroscopic information of the cell sample can be extracted from the reflected terahertz wave signal.

Cao et al. from Zhejiang University used ATR terahertz time-domain spectroscopy to characterize and differentiate human colorectal cancer epithelial cells [14]. Two human colorectal cancer epithelial cell lines (DLD-1 and HT-29) and a mouse fibroblast cell line (3T3) were used as research subjects. During each measurement, the cell solution was gently blown with a pipette to prevent the sample cells from settling

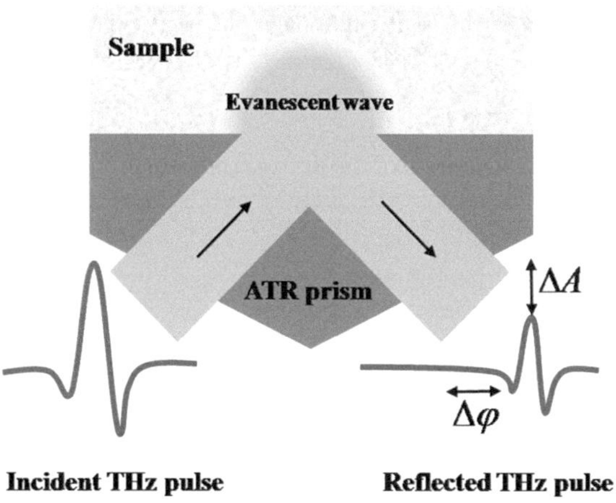

Fig. 6.15 Optical path of attenuated total reflection terahertz time-domain spectroscopy measurement. Reprinted with permission from Elsevier from Ref. [13], ©2017. Permission conveyed via RightsLink, licence no. 6079140954756, 1 June 2025

at the bottom. The terahertz spectra of the refractive index, absorption coefficient, and dielectric loss for the three cell lines were measured. The original terahertz spectra were affected by background noise from the external environment and interference from the terahertz time-domain spectroscopy system, which made the sample spectra lack authenticity and masked the true terahertz response of the cell samples. The Savitzky-Golay (S-G) convolution smoothing algorithm was used to process the measurement data. This algorithm uses local partial least squares fitting values to replace the center value of the smoothing window and uses polynomial fitting to estimate multiple signal points, thus obtaining the best fitting value for the smoothing point.

The terahertz refractive index spectrum reflects the biological characteristics of a cell's internal composition and molecular structure, playing a vital role in cell characterization. The original and processed refractive index spectra of human colorectal cancer epithelial cell lines (DLD-1 and HT-29) and mouse fibroblast cell line (3T3) solutions are displayed for concentrations increasing from 5×10^4 cells/ml to 5×10^6 cells/ml. For each cell line, the refractive index spectra of six concentrations exhibit distinct terahertz responses. There is no significant difference in the refractive index below 0.5 THz, whereas pronounced features appear at higher frequencies. All concentrations share a decreasing trend with frequency, yet the refractive index depends nonlinearly on concentration, and the magnitude of change varies among the three cell lines.

Principal Component Analysis (PCA) is a feature analysis and dimensionality-reduction technique that extracts low-dimensional principal components from the original data set by linearly recombining the original features. PCA compresses the data and highlights key information by identifying components that capture the distinctive characteristics of the six cell lines at different solution concentrations while eliminating redundancy. PCA is employed for feature extraction and cell characterization, and the Random Forest (RF) method is subsequently applied to differentiate cancer cell lines by analyzing the three feature parameters and comparing their results.

The distribution of the first two principal components extracted from the refractive index, absorption coefficient, and dielectric loss tangent of different-concentration 3T3 cell solutions is illustrated. The combined contributions of PC1 and PC2 for these three parameters reach 99.5%, 99.06%, and 99.35%, respectively—sufficient to represent the intrinsic information of 3T3 cells. Within the refractive-index panel, PC1 and PC2 obtained via PCA distinctly separate the concentration levels of 3T3 cell solutions. Clear differences are evident when describing the feature parameters across concentrations. Consequently, refractive index, absorption coefficient, and dielectric loss tangent enable discrimination among different solution concentrations of the same cell line through PCA.

Based on the above research and cell feature analysis, the Random Forest method can distinguish between human colorectal cancer cells and normal cells. During the classification process, both types of cancer cell lines are used as samples. The data is subjected to five-fold cross-validation, and the principal component analysis process can improve classification performance using any feature parameter. Since these three feature parameters have different sensitivities to material characterization, choosing one of them will lead to different classification results. It should be noted that the absorption coefficient has the best recognition effect on cancer cells. After principal component analysis detection, the recognition accuracy rate rises to 93.8%, the recall rate is 87.5%, and the F1 score is 89.1%. ATR terahertz time-domain spectroscopy has great potential in human cancer cell recognition and provides an experimental basis for liquid biopsy.

During the measurement process, to reduce the error caused by system phase inaccuracy, a double-path ATR terahertz time-domain spectroscopy can be used. The refractive index of high-resistance silicon in the terahertz band is about 3.415. A terahertz component including two high-resistance silicon attenuated total reflection prisms is designed for the terahertz time-domain spectroscopy system, which are fixed on the mechanical stage for reference signal measurement and sample measurement, as shown in Fig. 6.16. In this way, reference and sample measurements can be alternated by moving the terahertz beam, avoiding the long-term instability problem in traditional terahertz time-domain spectroscopy measurements.

The sample attenuated total reflection prism is designed with a medium reservoir that can hold about 3 ml, where about 3 million live cells can be attached for measurement. After observing the cells grow to 100% fusion with the upright microscope on top of the attenuated total reflection prism, the terahertz spectrum is obtained by measurement. With double-path attenuated total reflection terahertz time-domain spectroscopy, the oxidative stress response of live cells can be monitored [15]. When the cell suspension reaches a cell density of 1×10^6 cells/ml, 3 ml of the above cell suspension is inoculated onto the sample attenuated total reflection prism and cultured under conditions of 37°C and 5% CO_2. Before each experiment, the sample attenuated total reflection prism with cells is cleaned with

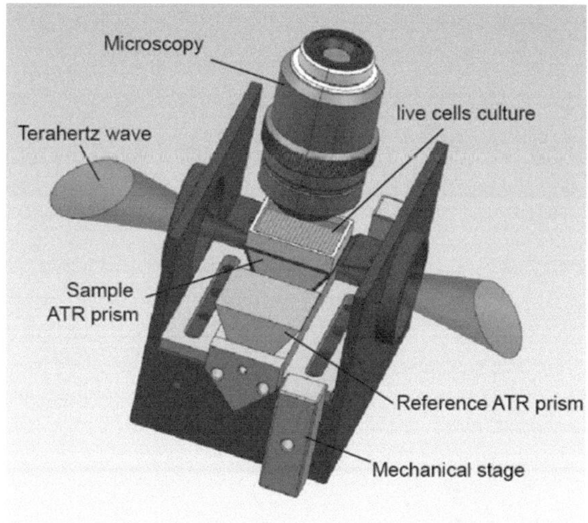

Fig. 6.16 Dual optical path attenuated total reflection terahertz time-domain spectroscopy measurement. Reprinted from Ref. [15], ©2018, under CC-BY 4.0

live cell imaging solution (LCIS), and relevant data is obtained using the sample attenuated total reflection prism and reference attenuated total reflection prism in the absence of cells. After measuring normal cells in standard live cell imaging solution, 3 μl of 10 M stock solution is added to the 3 ml cell culture on the sample attenuated total reflection prism, exposing the cells on the sample attenuated total reflection prism to 10 mM hydrogen peroxide. The cells are further cultured for 2 h, and the terahertz attenuation total reflection spectrum of the cells is detected every 6 min. The standard solution without cells is monitored by scratch cells and cleaned three times with live cell imaging solution. To optimize the terahertz spectrum measured from live cells and minimize interference from the culture medium, a homemade attenuated total reflection terahertz time-domain spectroscopy was used to measure human mammary epithelial cells MCF10A. To verify the accuracy of the double attenuated total reflection prism method, the live cell imaging solution was measured multiple times at different times. Compared with the traditional method that only performs reference measurements at the beginning, the error of the results obtained by this method is much smaller.

The time-domain terahertz pulses of live cells were measured at three different times, and then the average value was taken to determine the repeatability of the measurement results. Due to the effect of absorption, the average amplitude of MCF10A cell culture solution and live cell imaging solution is less than the average amplitude of the reference pulse. The average spectral amplitude of MCF10A cell culture solution is generally higher than that of live cell imaging solution, indicating that bulk water absorbs more terahertz waves than a single layer of live cells. The dielectric constant and dielectric loss of human mammary epithelial cells were quantitatively determined, and the complex dielectric constant of human mammary epithelial cells (MCF10A) in live cell imaging solution and the complex dielectric constant of live cell imaging solution were compared, as shown in Fig. 6.17. The

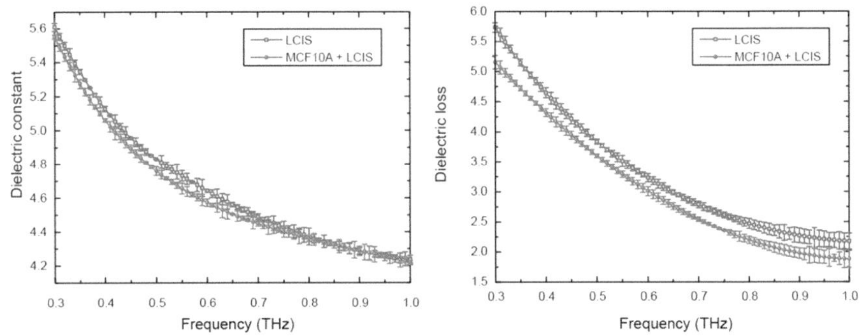

Fig. 6.17 Dielectric constant and dielectric loss of human breast epithelial cells (MCF10A). Reprinted from Ref. [15], ©2018, under CC-BY 4.0

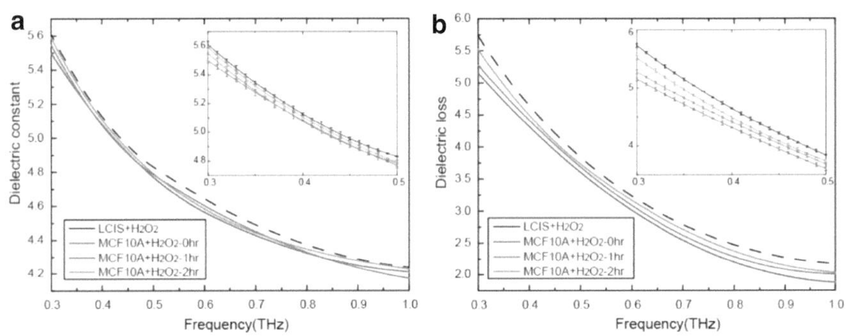

Fig. 6.18 Dielectric constant and dielectric loss of human breast epithelial cells (MCF10A) after addition of hydrogen peroxide. Reprinted from Ref. [15], ©2018, under CC-BY 4.0

results show that the dielectric constant of MCF10A cells has not changed, except for a slight decrease in the range below 0.5 THz. The dielectric loss of MCF10A cells has a significant downward trend from 0.3 THz to 0.9 THz. The trend of the complex dielectric constant of human mammary epithelial cells (MCF10A) below 1 THz is consistent with the results reported in colon cancer cells (DLD-1) [14].

In order to study the terahertz dielectric response of human mammary epithelial cells (MCF10A) under oxidative stress, a high concentration of hydrogen peroxide (10 mM) was added to the cells cultured on the attenuated total reflection prism. After adding hydrogen peroxide, it quickly degrades. After treatment with 0.1 mM hydrogen peroxide for one hour, only 1.6% of the cells remained viable. A concentration of 10 mM rapidly induces biomolecular responses and causes extensive cell death, making it suitable for short-term in situ monitoring of cells cultured under atmospheric conditions. The experimental results are shown in Fig. 6.18. After adding hydrogen peroxide, the dielectric constant did not change, but the dielectric loss increased.

In order to study the relationship between the death of MCF10A cells measured by flow cytometry and the terahertz dielectric response of cells after hydrogen peroxide treatment, the time-domain peak amplitude of the time-domain terahertz pulse is calculated and the peak amplitude of the terahertz time-domain signal is selected to calculate the contrast. By analyzing the contrast, it slowly grows within the first hour, consistent with the slow increase in cell death rate. In contrast, in 1–2 h, the terahertz contrast increases, and the cell death rate also significantly increases. Cells are more likely to undergo apoptosis in the first hour, so the permeability of the cell membrane has a smaller impact. As more and more cells die, late-stage apoptotic and necrotic cells undergo cytoplasmic leakage, and the integrity of the plasma and nuclear membranes also decreases, leading to a rapid increase in terahertz contrast. The dual optical path attenuated total reflection method can be used to monitor cell activity during oxidative stress, without the need for additional labels, and this method can also be used for research on drug antioxidant properties.

6.3.3 Terahertz Biosensor Detection of Cells Using Metamaterials

Metamaterials are a new type of artificial composite electromagnetic material that has rapidly developed over the past decade or so. By utilizing the oscillation of periodic resonant units under the excitation of electromagnetic waves, they exhibit unusual electromagnetic properties not found in natural materials, such as negative permittivity, negative magnetic conductivity, and negative refractive index. Because terahertz waves have longer wavelengths than infrared and visible light, the sub-wavelength microstructure units of terahertz metamaterial devices are on the μm scale, and their fabrication can utilize mature semiconductor manufacturing processes, making device integration and miniaturization easy. Based on these characteristics, terahertz metamaterials can be applied in many areas of current scientific research, providing new methods and means for cell sensing and detection research.

Jin et al. from Nanjing University designed a biosensor based on terahertz metamaterials, which was used for label-free measurement of cell apoptosis [16]. This terahertz metamaterial sensor, composed of a planar array of photo-etched concentric sub-wavelength metal ring resonators on a 10 μm thick polyimide (PI) substrate, has high sensitivity. The metamaterial was measured under normal incidence using a transmission-type terahertz time-domain spectroscopy system. The corresponding transmission spectrum. Due to the resonance of two adjacent rings, four resonance peaks appear, similar to the phenomenon of electromagnetically induced transparency.

In the experiment, a single layer of oral squamous cell carcinoma cells (HSC3) and normal epithelial cells (HaCaT) were selected as research subjects to evaluate sensor performance, using terahertz time-domain spectroscopy to measure different concentrations of single-cell suspensions. The difference in resonance frequency when the metamaterial is present and absent of cell analytes is defined as the relative frequency shift Δf. The relative frequency shift Δf increases with the increase in cell concentration, but reaches saturation at high concentrations. Cancer cells HSC3 and normal cells HaCaT have similar characteristics, which is because the cell layer changes the dielectric environment of the metamaterial sensor.

This terahertz metamaterial sensor was used to conduct research on cell apoptosis detection, studying the relationship between the relative change in resonance frequency and the rate of cell apoptosis. Changes in the number of cells on the metamaterial alter the effective dielectric constant of the dielectric environment above the biosensor. Flow cytometry analysis revealed a linear relationship between the relative shifts in the four resonance frequencies and the apoptosis rate.

Yao et al. applied metamaterials to cell sensing research and proposed a new type of terahertz metamaterial biosensor based on electromagnetically induced transparency [17], as shown in Fig. 6.19. They used a symmetrically broken double-aperture ring resonator to achieve electromagnetically induced transparency plasmonic resonance, creating a transparent window at 1.67 THz. The positions of these two apertures are asymmetrical, introducing a symmetry break.

The sample preparation and cell culture process is shown in Fig. 6.20. First, a polyimide layer is spin-coated onto a silicon substrate, then a double-aperture ring structure is obtained using standard photolithography. After that, a 20 nm layer of titanium and a 200 nm layer of gold are deposited, and the polyimide layer with the metal double-aperture ring is peeled off from the silicon substrate. Finally, oral cancer cells HSC3 are seeded and cultured on the surface of this structure, and the

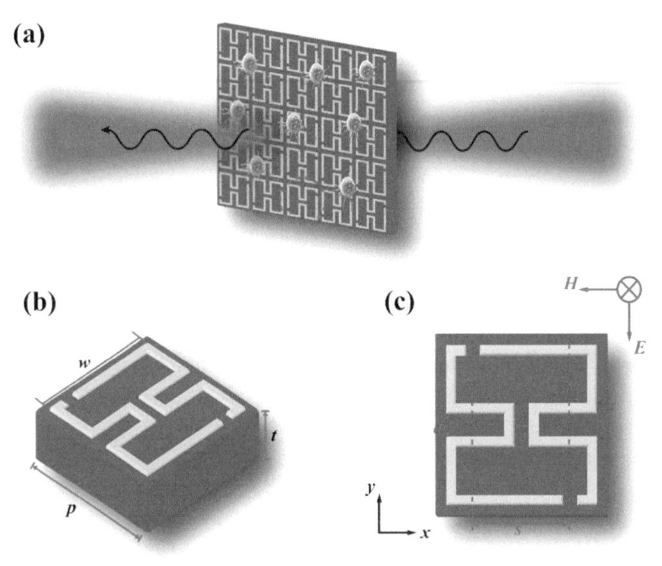

Fig. 6.19 Schematic diagram of double split-ring terahertz metamaterial structure. Reprinted from Ref. [17], ©2019, under CC-BY 4.0

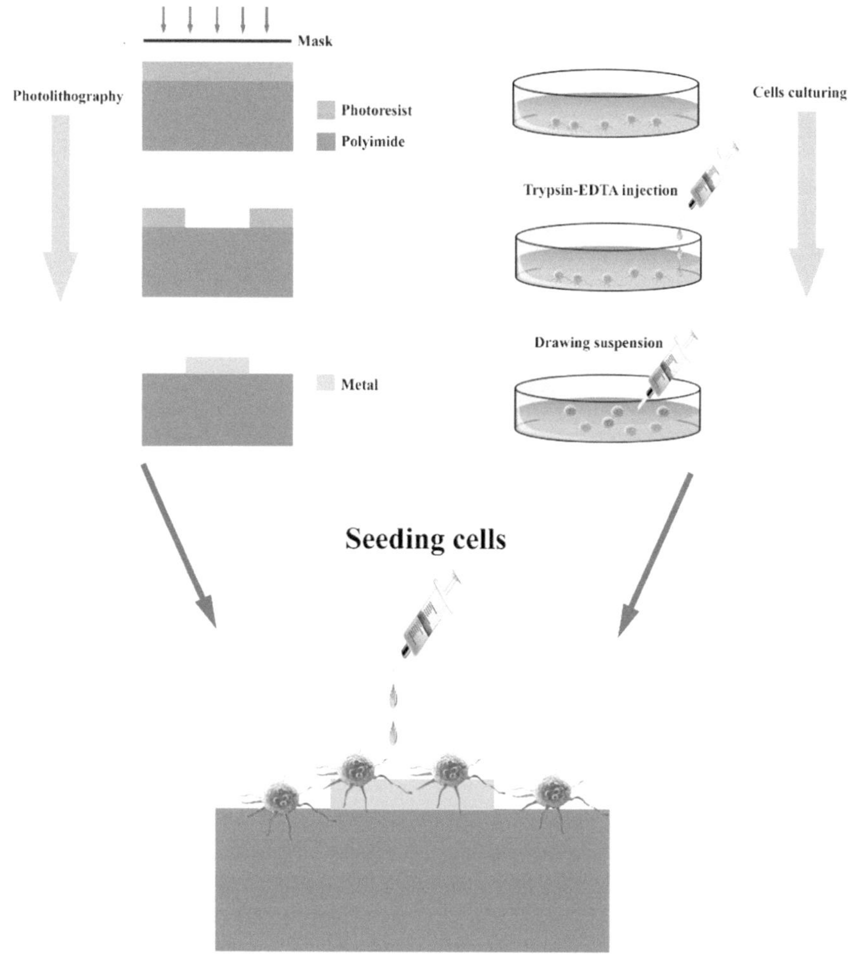

Fig. 6.20 Schematic diagram of the process of culturing cells on terahertz metamaterial biosensor, including photolithography preparation steps and cell culture process. Reprinted from Ref. [17], ©2019, under CC-BY 4.0

resonance characteristics of this terahertz metamaterial sensor are measured using a terahertz time-domain spectroscopy system.

Simulation and experimental measurements were used to study the resonance mechanism of the terahertz metamaterial sensor structure, and the terahertz spectrum is shown in Fig. 6.21a. Apart from a slight difference in the intensity drop at 1.15 THz and 1.87 THz, the experimental results match the simulation results quite well. This may be due to differences between the simulation conditions used and the manufacturing errors in the actual experiment. A transmission peak with an efficiency of 55% appears near 1.67 THz, and the absorption at this frequency drops to 30%. Due to this weak transparency phenomenon, some mode coupling effects exist in the designed double-aperture ring structure. To further study its physical mechanism, they analyzed the electric field, z-axis magnetic field, and surface charge density at three different frequencies, as shown in Fig. 6.21b–g. For 1.2 THz, the transmission will be almost zero. The electric field distribution shown in Fig. 6.21b indicates that an electric dipole resonance mode is generated at this frequency. At the same time, incomplete semi-circular currents are formed on both sides of the metal structure, marked with a black dashed circle. This quasi-circular surface current induces a weak magnetic field located only at the two ends of the metal ring, as shown in Fig. 6.21c. When the frequency increases to 1.67 THz, forming a pair of electric quadrupoles, as shown in Fig. 6.21d. At the same time, two complete circulating currents flow along the upper and lower loops, as marked by the black dashed circles in Fig. 6.21e. Due to this circulating current along the electric quadrupoles, an anti-phase magnetic field is generated in the space surrounded by the metal ring, and its intensity increases, thereby generating magnetic dipole resonance. From the above analysis, it can be found that the electric field distribution is converted into a quadrupole resonance mode, that is, non-radiative damping. A possible explanation for this change in resonance mode is the coupling of the magnetic dipole/electric quadrupole resonance (dark mode) with the electric dipole (bright mode) originating from 1.20 THz, destructive interference suppresses the bright mode, and energy is coupled to the bright mode, under which the quadrupole resonance dominates. Due to the suppression of the electric dipole, radiation loss is greatly reduced, and transmission performance is correspondingly improved. This can be confirmed by the black line in Fig. 6.21a, where transmission increases by 55% at 1.67 THz, and absorption is reduced to 30%. In the region around 2.0 THz, the electric quadrupole is again converted into a dipole mode, as the circulating current is incomplete, the anti-phase magnetic field is not located inside the metal ring, as shown by the black dashed circle in Fig. 6.21g. This phenomenon indicates that interference decreases again as the frequency increases. In this case, radiation loss dominates, and the transmission window disappears. The above analysis shows that different types of resonances have different electric distributions at the split positions. For example, at 1.20 THz, the electric field is distributed in the same direction at both ends of the split. But when the frequency is increased to 1.67 THz, the electric field becomes a reverse phase distribution. The split position is related to the nature of electromagnetic induction transparency resonance.

To investigate the sensitivity of metamaterials exhibiting electromagnetically induced transparency for biosensing applications, oral cancer cells (HSC-3) were selected as analytes and cultured on the surface of the metamaterial. In the experiment, cell concentration changes were monitored by measuring the corresponding terahertz spectra of the metamaterial biosensor. As the cell concentration increases (7×10^5 cells/ml), the transmittance decreases. Under the action of anticancer drugs, more cells are inactivated, the cell concentration decreases, and a lower transmittance is detected. Normally, the transmittance of each sample is different. Therefore, most research does not evaluate the performance of biosensors by analyzing the transmission intensity, but by the resonance frequency shift. For cells of different concentrations, the resonance frequency of the designed structure starts to redshift from

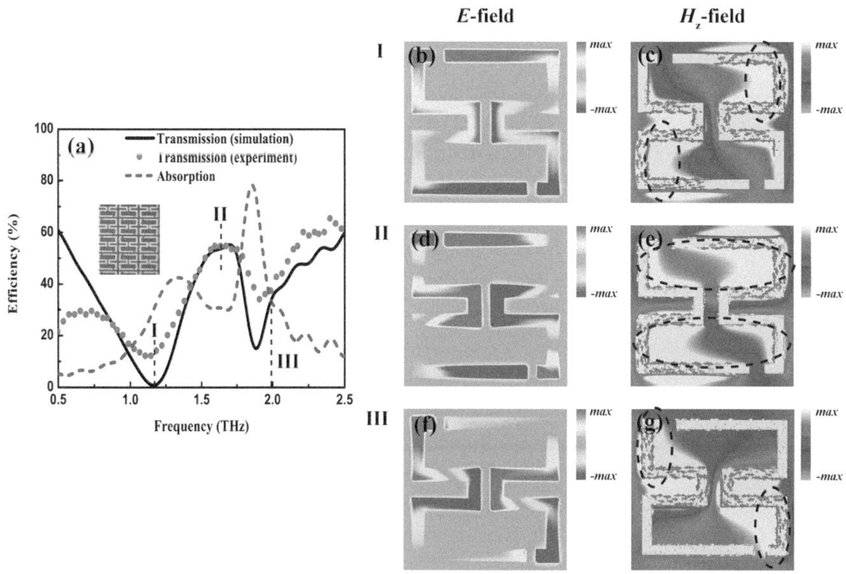

Fig. 6.21 **a** Simulated and experimentally measured transmission curve and absorption curve; **b–g** electric field, magnetic field, and surface current distribution in different frequency ranges. Reprinted from Ref. [17], ©2019, under CC-BY 4.0

1.65 THz. To more clearly show this change, the frequency shift Δf under different cell concentrations. The measurement results show that when the cell concentration increases from 1×10^5 cell/ml to 7×10^5 cell/ml, the resonance peak shift Δf increases from 50 to 90 GHz. The resonance peak of electromagnetic induction transparency is very sensitive to changes in cell concentration. In addition, as the cell concentration increases, the slope of Δf approaches flat, indicating that the sensitivity is limited in actual situations. When the cell concentration is 7×10^5 cell/ml, the maximum sensitivity obtained in the experiment is close to 900 kHz/cell/ml, which means that each cell per milliliter will cause a frequency shift of 900 kHz. This feature helps to establish the relationship between cell concentration and resonance peak frequency shift, and provides another method for studying cancer cell activity from a physical perspective.

Among various cell activities, cell apoptosis is one of the important features for evaluating the evolution of cancer cells. In order to demonstrate the application prospects of their designed biosensor based on EIT-like metamaterials in cancer cell apoptosis detection, the transmittance of the biosensor under different concentrations of the anticancer drug cisplatin and different drug action times was measured. Before using cisplatin, cells with an original concentration of 2×10^6 cells/ml were cultured on the surface of the metamaterial biosensor. These samples were divided into two groups and treated under different conditions. In the experiment, the action time was fixed at 24 h, the cisplatin concentration was 1–15 μM, and the cisplatin concentration was fixed at 1 μM, and the drug action time was 24 h to 72 h. The experimental results show that as the concentration of cisplatin increases, the resonance peak shows a relative blueshift, and the frequency shift Δf decreases from 140 to 70 GHz. Combined with the above analysis, it shows that the inactivation of cells by anticancer drugs leads to a decrease in cell concentration. In other words, the higher the concentration of cisplatin, the more it will promote cell apoptosis. This result is consistent with the measurement using the CCK-8 reagent kit. In addition, when the concentration of cisplatin increases from 1 μM to 15 μM, cell apoptosis accelerates from 5.3% to 35.4%, and the half-width increases from 237.9 GHz to 305.4 GHz. When the drug action time increases from 24 to 72 h, the resonance peak shows a blueshift, indicating a decrease in cell concentration. The frequency shift Δf From 140 GHz down to 40 GHz, with the prolongation of drug action time, cancer cells are effectively inactivated, and cell apoptosis accelerates. The half-width at half-maximum (HWHM) increases from 237.8 GHz to 337.6 GHz, and cell apoptosis increases from 5.3% to 14.5%. Therefore, based on the above analysis of cancer cell apoptosis, it is shown that the frequency shift Δf and HWHM detected by the metamaterial biosensor are consistent with the results obtained by biological measurement methods. This proves that terahertz metamaterial biosensors have great potential in cancer cell detection applications and provide a new, fast, and low-cost biosensor detection scheme from a physics perspective.

Zhu and Chen et al. used an electromagnetically induced transparent terahertz metamaterial sensor for label-free biosensing of glioma cell molecular classification [18]. As shown in Fig. 6.22a–c, this electromagnetically induced transparent metamaterial consists of metal cut wires and split-ring resonators, with 100 unit structures in both x and y directions. The entire physical size of the metal array is 5 mm × 5 mm, capable of producing a polarization-independent transparent window at 2.24 THz. Two types of glioma cells (mutant and wild type) were cultured and adhered to the metal surface of the proposed metamaterial biosensor. Figure 6.22d–f show the microscopic images of the biosensors cultured with cell concentrations of 20×10^4 cells/ml, 40×10^4 cells/ml and 80×10^4 cells/ml. Before measurement, the prepared biosensors were washed on both sides with phosphate-buffered saline and dried. The terahertz transmission spectrum of the terahertz metamaterial sensor was measured using a terahertz time-domain spectroscopy system. After cell culture and preparation, the biosensor was fixed on the sample holder in the terahertz time-domain spectroscopy system, perpendicular to the line between the emitter and detector. The terahertz transmission spectrum of the biosensor covered with glioma cells was detected in an environment with a temperature of 26°C and a relative humidity of 3%. The time-domain data was converted into frequency-domain data based on Fourier transform, and then the transmission data was normalized with air reference values to obtain the frequency and amplitude responses of cells at different concentrations.

To study the sensitivity of the metamaterial sensor, the refractive index sensing performance was first simulated numerically. In the simulation, a cell layer with a thickness of 10 μm was added to the terahertz metamaterial sensor. Figure 6.23a shows that when the refractive index of the cells increases from 1.1 to 1.9, the transmission peak frequency significantly redshifts, and the amplitude change is not

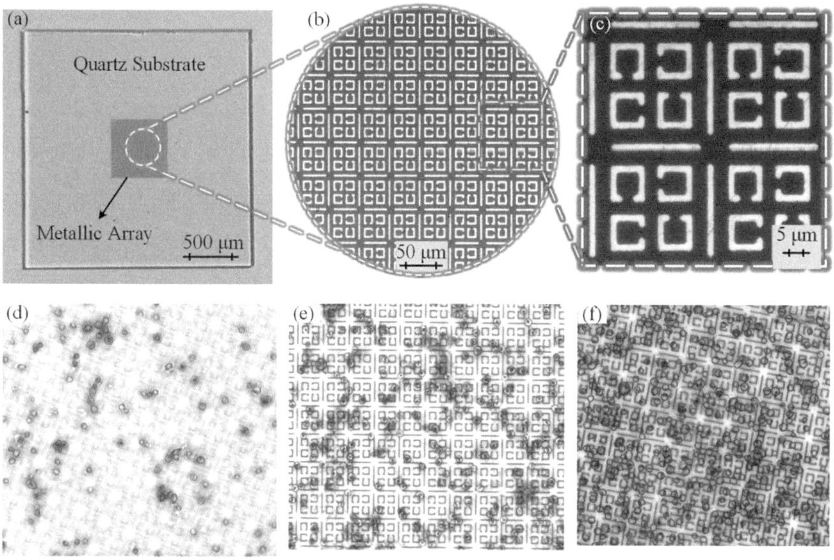

Fig. 6.22 **a–c** Physical and structural diagrams of terahertz metamaterial sensors **d–f** Glioma cells with concentrations of 20 × 10^4 cells/ml, 40 × 10^4 cells/ml, and 80 × 10^4 cells/ml attached to the sensor. Reprinted with permission from Elsevier from Ref. [18], ©2021. Permission conveyed via RightsLink, licence no. 607924106993, 1 Aug. 2025

significant. The working frequency of the electromagnetic induction transparency efficiency is determined by the resonant structure and the surrounding medium. When a high refractive index analyte is covered on the upper surface of the metal resonator, the equivalent dielectric constant of the surrounding medium increases accordingly, leading to changes in the induced electric field, ultimately changing the central resonant peak. The frequency shift Δf and absolute peak transmission rate corresponding to different refractive indices are shown in Fig. 6.23b, showing an approximate linear relationship. In addition, due to the high quality factor Q of the electromagnetic induction transparency effect, the theoretical sensitivity of this biosensor is $S = \Delta f/\Delta n \approx 496.01$ GHz/RIU.

Further simulation studies the effect of the thickness of the test cell layer on terahertz metamaterial biosensing. Figures 6.23c, d show the movement trend of the terahertz spectrum under different thicknesses of the analyte when the refractive index is 1.6. It can be seen that this terahertz metamaterial biosensor is sensitive to the thickness of the test cells, and the frequency shift Δf increases with the increase in thickness. In addition, the relationship between Δf and thickness shows that when the thickness increases from 1 μm to 8 μm, the frequency of the transmission peak value changes significantly, and when the thickness reaches 14 μm, it remains unchanged. This is because the excitation field coupling effect of the metal structure is confined to its surface area.

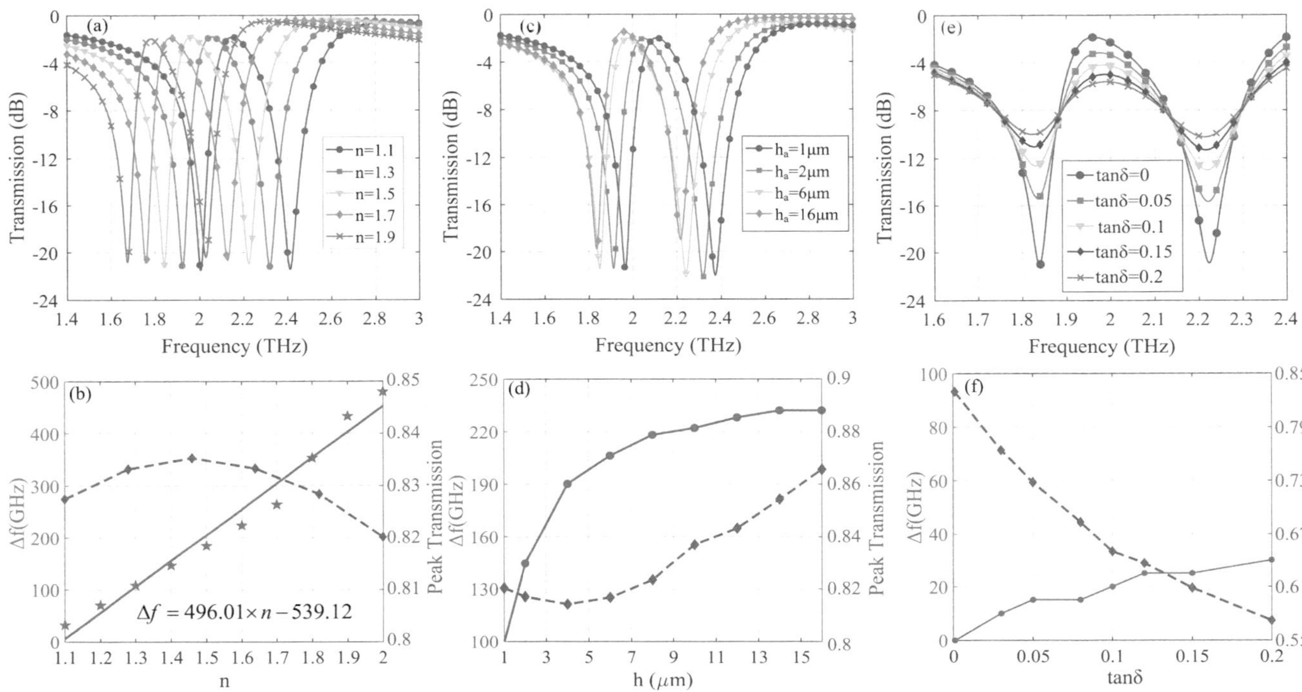

Fig. 6.23 Simulation results of the characteristics of cells to be tested with metamaterial biosensors **a–b** Terahertz spectra with different refractive indices **c–d** Terahertz spectra with different thicknesses **e–f** Terahertz spectra with different dielectric loss tangents. Reprinted with permission from Elsevier from Ref. [18], ©2021. Permission conveyed via RightsLink, licence no. 607924106993, 1 Aug. 2025

The dielectric loss of the test cells also affects the class electromagnetic induction transparent transmission spectrum of the proposed biosensor. In the simulation, the loss tangent (tanδ) is used to characterize the dielectric loss of the test cells. As can be seen from Fig. 6.23e, f, under a fixed refractive index of 1.6 and a thickness of 10 μm, when the loss tangent increases from 0 to 0.2, the frequency shift Δf undergoes a slight change, while the peak transmission amplitude decreases from 0.83 to 0.57. Compared with other situations represented in Fig. 6.23b, d, f, the dielectric loss is the most sensitive factor influencing the amplitude of the electromagnetically induced transparency-like transmission peak. This characteristic provides a new degree of freedom for the application of biomolecular detection.

Experimental research has verified the actual performance of terahertz metamaterials for biosensing. Two types of glioma cells were cultured on the surface of the metamaterial. The ability of the proposed biosensor to detect changes in cell concentration was first studied. Figure 6.24a, b present the terahertz spectra of biosensors covered with glioma cells of different concentrations, either mutant or wild type, with the terahertz spectrum of a cell-free sensor provided for comparison. The transmission spectra of both types of glioma cells show significant changes in resonance frequency and peak amplitude. For instance, when the concentration of mutant glioma cells increases from 2×10^5 cells/ml to 8×10^5 cells/ml, the resonance frequency shifts from 2.274 THz to 2.228 THz, and the peak amplitude slightly decreases from 36.73% to 34.21%. A similar trend is observed in wild-type glioma cells. The actual sensitivity in the experiment is defined as $S = \Delta f/\Delta c$, where Δc represents the change in cell concentration. The calculated maximum sensitivities for mutant and wild-type glioma cells are 122 kHz/cell/ml and 248.75 kHz/cell/ml, respectively. These results suggest that the proposed electromagnetic induction transparent metamaterial is an effective platform for high-sensitivity detection of glioma cells from a physical perspective. The discrepancy between the simulated sensitivity and the measured sensitivity is due to slight changes in the refractive index of the cells and noise in the measurement environment.

The ability of the terahertz metamaterial biosensor to distinguish between different types of cells was further validated experimentally. Figure 6.24c shows the terahertz spectra of biosensors covered with mutant and wild-type glioma cells at the same concentration of 4×10^5 cells/ml. There is a significant difference in the resonance frequencies of these two types of glioma cells. The measurement results indicate that the biosensor can effectively identify cancer cells at specific cell concentrations. However, in actual detection, cultivating samples with specific cell concentrations is time-consuming. To effectively identify the molecular types of glioma cells, the frequency shift and peak changes were evaluated, revealing a dependency on the concentration of mutant and wild-type glioma cells as shown in Fig. 6.24d. It can be seen that for wild-type glioma cells, changes in cell concentration cause a resonance frequency shift close to 0.2 THz, while the frequency shift for mutant glioma cells is only 0.12 THz, with changes in concentration mainly reflected in the change in peak amplitude. The two lines in Fig. 6.24d separate, indicating the areas of frequency and amplitude changes for mutant cells (blue area) and wild-type glioma cells (yellow area). This clearly shows that once the two degrees of change are

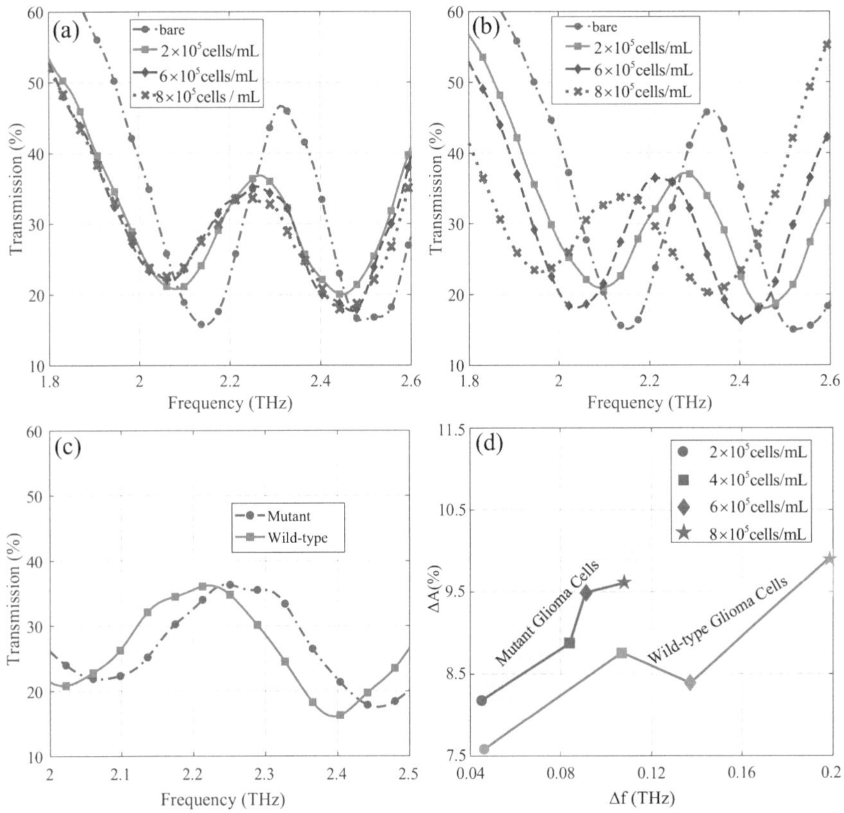

Fig. 6.24 Terahertz transmission spectra of metamaterial biosensors **a** Mutant IDH glioma cells and **b** Wild-type IDH glioma cells **c** Transmission spectra of both cell types at a concentration of 4×10^5 cells/ml **d** Resonant frequency and peak amplitude changes with concentration for mutant and wild-type glioma cells. Reprinted with permission from Elsevier from Ref. [18], ©2021. Permission conveyed via RightsLink, licence no. 607924106993, 1 Aug. 2025

measured, the type and concentration of glioma cells can be directly read from Fig. 6.24d. It is worth noting that the two types of glioma cells have different water content and nuclear sizes, leading to differences in refractive index and dielectric loss. The changes in resonance frequency and amplitude of the designed metamaterial sensor are highly related to the refractive index and dielectric loss of the sample to be analyzed. The experimental results show that glioma cell classification has great potential in practical applications, providing a new approach for rapid, low-cost detection of different types of cancer cells.

6.4 Biological Tissue Detection

Tissue, in biology, is a level between cells and organs. In multicellular organisms, cell differentiation results in different cell groups, each of which is composed of many morphologically similar cells with the same structure and function, along with intercellular substances. These cell groups are called tissues. Different tissues work together to form organs. To maintain cell and tissue morphology during terahertz spectroscopy experiments and storage, and to prevent tissue autolysis and necrosis, tissue sample fixation is crucial. Tissue sample fixation includes embedding in paraffin or freezing. Embedding is important for maintaining tissue morphology and providing tissue support during slicing, typically into thin sections (5–10 μm) for further study. When creating paraffin-embedded tissue samples, the tissue must first be fixed, then embedded in paraffin. After slicing, fixation is achieved immediately through perfusion or immersion. This process usually takes 4–24 h. Frozen tissues can be prepared by immersing the tissue in liquid nitrogen, isopentane, or embedding the sample in dry ice. Frozen tissues are cut using a cryostat. The obtained slices can be stored at -80 °C for up to a year. Alcohols such as methanol or ethanol can usually be used to fix frozen tissue slices.

Cerebral ischemia is one of the common diseases in neurosurgery, which can lead to blurred consciousness in patients, and prolonged cerebral ischemia can even cause permanent damage to brain tissue. Existing Magnetic Resonance Imaging (MRI) and Computed Tomography (CT) imaging technologies can be used for preoperative detection of cerebral ischemia, but they have issues such as complex equipment operation, long duration, and high cost, and cannot detect ischemic areas during surgery. Additionally, intraoperative cerebral ischemia can be detected by monitoring local cerebral blood flow and brain electrical activity, but this method has problems with low time and spatial resolution and low accuracy. Therefore, there is a need to develop a fast, accurate, economical, and harmless method for detecting cerebral ischemia.

Cerebral ischemia is actually a change in the water content of brain tissue, and terahertz waves are very sensitive to differences in water content. The absorption coefficient of the terahertz spectrum is first considered to characterize the degree of ischemia. In the experiment, adult healthy rat brain tissue was selected as the research object, and the transmission terahertz time-domain spectra of normal brain tissue, mild ischemic brain tissue, moderate ischemic brain tissue, and severe ischemic brain tissue were measured respectively. The results are shown in Fig. 6.25. For the same sample, different measurement points are selected, and the error bars representing the standard deviation of the average value represent the terahertz absorption coefficient of the tissue. The results show that within the measurement range of 0.7–2.6 THz, the average absorption coefficient increases monotonically with the terahertz frequency. At the same terahertz frequency, the terahertz spectral absorption coefficient of rat brain tissue increases with the severity of brain injury [19].

Different degrees of cerebral ischemic tissue can be qualitatively distinguished through terahertz time-domain spectroscopy. In clinical surgery, it is particularly

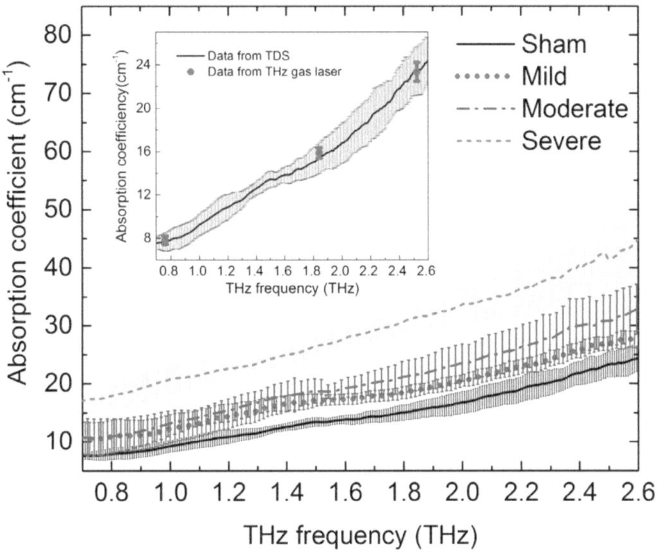

Fig. 6.25 Terahertz absorption spectra of rat brain tissue with different degrees of ischemia. Reprinted from Ref. [20], ©2018, under CC-BY 4.0

important to identify the boundaries of ischemic brain tissue, and there is currently a lack of precise positioning methods for intraoperative ischemic brain tissue. Using terahertz imaging technology to locate ischemic brain tissue has become the goal of the next step of research exploration. Figure 6.26a shows that the brightness of the MRI image of the high signal area of ischemic brain tissue is higher than that of the normal area, indicating that the water content of ischemic brain tissue is higher than that of normal tissue. In addition, different degrees of high signals appear at the injured site, which matches well with the order of injury. After MRI scanning, the obtained fresh specimens are trimmed with a slicer at −2 °C until the coronal surface matches the MRI image, and cut into uniform slices for terahertz transmittance imaging measurement. The terahertz intensity transmission image is shown in Fig. 6.26c, where lower transmittance indicates higher absorption by the sample. It can be clearly seen that the terahertz transmittance image of normal rat brain tissue looks very uniform, with an average transmittance of 23%. The terahertz images of different degrees of cerebral ischemic tissue are significantly different from normal tissue. The average transmittance values of the abnormal areas of mild, moderate, and severe cerebral ischemic tissues are about 20%, 19%, and 19% respectively, lower than normal brain tissue. The volumes of the abnormal areas of mild, moderate, and severe cerebral ischemic tissues gradually increase, and the positions of the ischemic areas in the terahertz images are similar to the corresponding MRI images. As the water content in brain tissue increases with the severity of cerebral ischemia, the obtained MRI and terahertz images show that water can be used as the main reagent

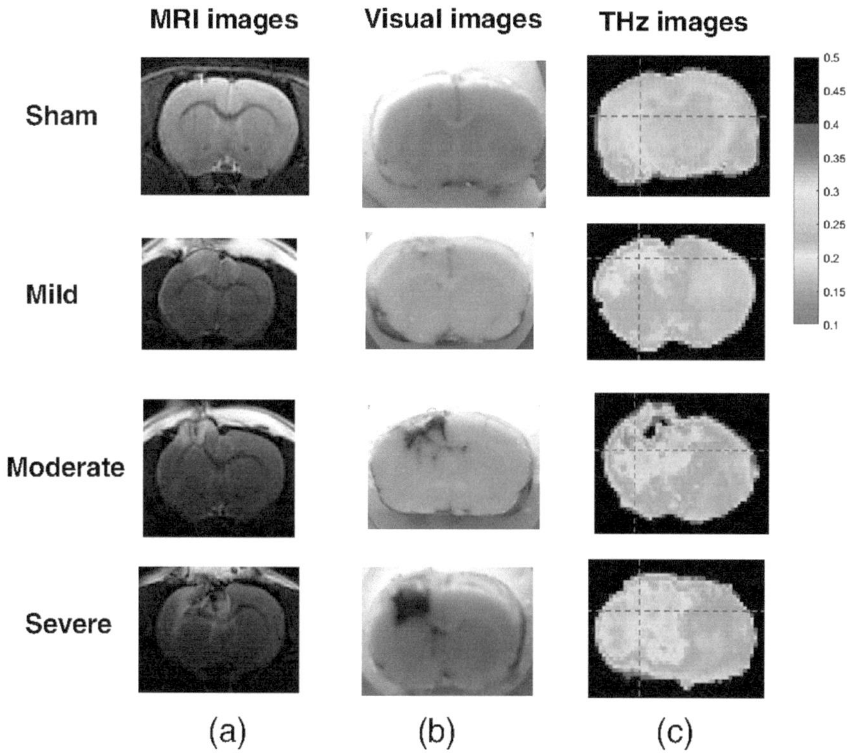

Fig. 6.26 Imaging comparison of rat brain tissue with different degrees of ischemia **a** Magnetic resonance imaging **b** Visual image **c** Terahertz imaging. Reprinted from Ref. [20], ©2018, under CC-BY 4.0

for terahertz cerebral ischemia imaging. The results also found that the smaller terahertz image contrast differences in normal tissue correspond to the gray matter and white matter areas, as shown in the visual image 6.26(b). In the visual image, the areas where the tissue color is white, the transmittance of terahertz waves is slightly higher. This is because the myelin in white matter is mainly composed of lipids, which have a lower absorption rate for terahertz waves than water. The distribution of white matter in the terahertz image corresponds to the corresponding T2-weighted MRI image [20].

Brain metastases refer to the spread of cancer cells from their original location to the brain, and are one of the most common brain tumors in humans. Figure 6.27a shows the sagittal T1-weighted MRI image of a metastatic brain tumor in a 57-year-old female patient. The patient has a history of lung tumor resection, and the tumor sample was collected in the area where the tumor cells are gathered next to the normal brain tissue. Metastatic brain tumors cannot be visually distinguished from normal tissue. The terahertz spectroscopy results are shown in Fig. 6.27b, and there are differences in the transmission spectra of normal tissue and brain metastatic

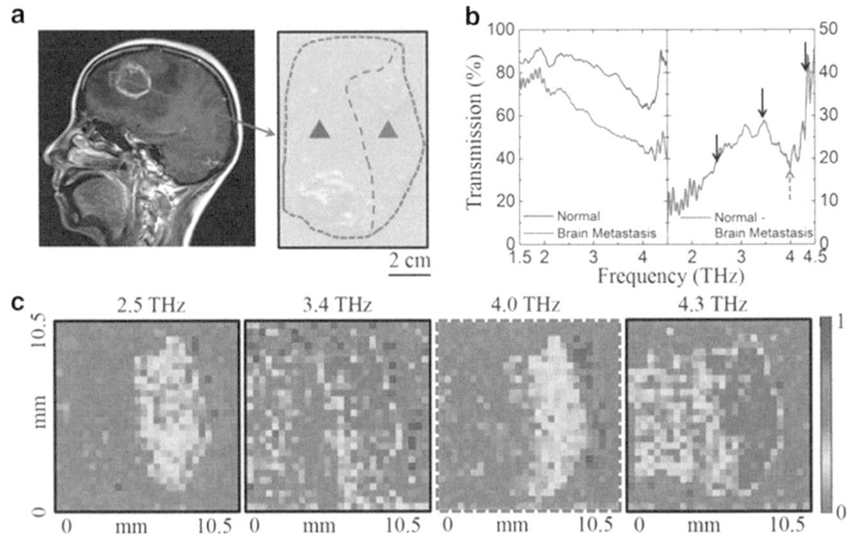

Fig. 6.27 Terahertz spectra and imaging results of human brain metastases **a** Human brain metastasis tissue sample **b** Terahertz spectra of human brain metastasis tissue and normal tissue **c** Terahertz imaging of human brain metastasis tissue and normal tissue at different frequencies. Reprinted from Ref. [21], ©2018, under CC-BY 4.0

tumor tissue. Subtracting the transmission spectrum of human brain metastatic tumor tissue and normal tissue, there are significant differences in the differential terahertz spectrum, where the distinction is most obvious at 3.4 THz and 4.3 THz. In order to achieve the boundary recognition of human brain metastatic tumors, the terahertz imaging of human brain metastatic tumor tissue at 2.5 THz, 3.4 THz, 4.0 THz, and 4.3 THz is analyzed. The highest contrast was observed at 3.4 THz and 4.3 THz, which is consistent with the terahertz spectral data obtained from terahertz time-domain spectroscopy measurements [21]. Although the strong absorption of water limits the penetration depth of terahertz waves, hindering the application of terahertz imaging technology in vivo, future higher power terahertz sources can alleviate this problem, making terahertz spectral imaging a superior technology for clinical detection.

6.5 COVID-19 Detection

Coronavirus disease 2019 (COVID-19) is an infectious disease caused by severe acute respiratory syndrome coronavirus 2 (SARS-CoV-2), leading to an ongoing pandemic and becoming one of the deadliest epidemics in human history. The World Health Organization (WHO) declared the COVID-19 outbreak a Public Health Emergency of International Concern, making the COVID-19 pandemic the most severe crisis faced by humanity since World War II.

The main methods for detecting the novel coronavirus currently include conventional polymerase chain reaction (PCR) testing and rapid antigen testing. PCR testing requires amplification of complementary deoxyribonucleic acid (DNA) and takes several hours to produce results, making it slow. Although rapid antigen testing can produce results within 20 min, the antigen–antibody interaction of SARS-CoV-2 has a high structural similarity with antigens within the same genus, resulting in low accuracy.

In January 2022, Lee et al. reported their results, using terahertz metamaterials to detect and identify the receptor binding domain of SARS-CoV-2 spike protein, quickly and accurately identifying the variant strains of COVID-19 [22]. Because terahertz waves have a low probability of interacting with biological samples and there is no technology to amplify the signal, it is difficult to directly detect live samples. To overcome this drawback, a biomolecular diagnostic platform was developed using a metamaterial sensor that amplifies the terahertz wave signal at specific bands. This platform achieves the amplification and identification of a very small amount of terahertz signals. Figure 6.28a introduces the schematic diagram of using terahertz metamaterial sensors to detect and distinguish between SARS-CoV-2 and SARS-CoV, and Fig. 6.28a introduces the AA sequence in the receptor binding domain (RBD) of SARS-CoV-2 and SARS-CoV. The yellow area represents SCRBD-1 and SC2RBD-1, and the blue area represents SCRBD-2 and SC2RBD-2. Figure 6.28c introduces the crystal structures of SCRBD-1,2 and SC2RBD-1,2.

Terahertz waves have a very wide band and can be used for spectroscopic analysis. They are also sensitive to the natural vibrations of biomolecules and can interpret unique information and differences in biological samples such as DNA and amino acids. Amino acids with excellent absorption capabilities in the terahertz band use terahertz metamaterial sensors to amplify the signal. The receptor binding domain samples of the spike protein are dispersed on the surface of the metamaterial, and the changes in the terahertz signal are observed, as shown in Fig. 6.29a. The corresponding terahertz time-domain signal is detected by terahertz time-domain spectroscopy and a fast Fourier transform is performed to obtain the corresponding terahertz spectral response. Terahertz metamaterial sensors with three different frequencies are used, with resonance frequencies at 1.16, 1.64, and 2.07 THz. The RBD of the spike protein in SARS-CoV-2 is a key component for the virus to enter, and compared to other structural proteins, it has a unique sequence and is considered an important diagnostic target. The receptor binding domains of the spike proteins of SARS-CoV-2 and SARS-CoV have highly similar secondary structures. Using metamaterial terahertz sensors with resonance frequencies of 1.16 THz, 1.64 THz, and 2.07 THz, the terahertz spectral fingerprints of the receptor binding domain of the spike protein are measured, as shown in Fig. 6.29b, c. These fingerprints originate from various structural features, including the peptide itself, single atom absorption spectra, and hydrogen bonds.

The changes and standard deviations of the terahertz frequency offset are analyzed based on the hydrophilicity and net charge of the receptor binding domain. The terahertz spectral characteristics of the spike protein receptor binding domain are evaluated through variations in frequency offset and the relative transmittance change

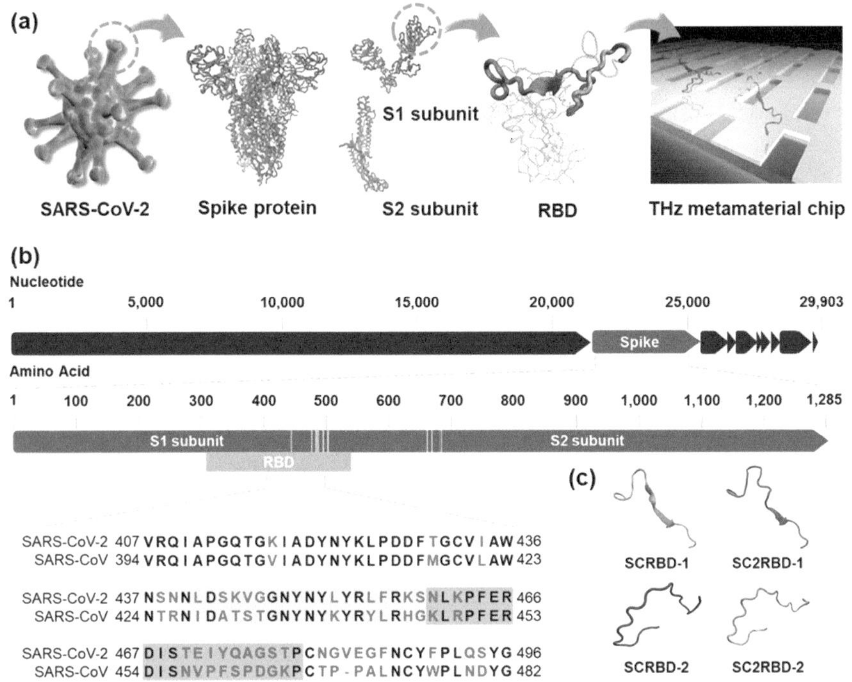

Fig. 6.28 **a** Schematic diagram of using terahertz metamaterials to detect and distinguish SARS-CoV-2 and SARS-CoV **b** Whole genome and AA sequence characteristics of SARS-CoV-2 **c** Crystal structures of SCRBD-1,2 and SC2RBD-1,2. Reprinted from Ref. [22], ©2022, under CC-BY 4.0

($\Delta T/T_0$) of the terahertz spectrum. Figure 6.30 shows the terahertz spectral response characteristics of two receptor binding domains of SARS-CoV-2 and SARS-CoV. For SCRBD-1 and SC2RBD-1, which have similar binary structures (partially α-helix and β-sheet), hydrophilicity and positively charged peptides lead to more effective molecular capture and higher membrane uniformity. The spectral differences at the resonant frequencies of 1.64 THz and 2.07 THz are attributed to the number of primary atomic absorption spectra, which have strong inherent absorption peaks at the measured frequencies. On the other hand, for SCRBD-2 and SC2RBD-2, which have κ-helices, significant signal differences are observed. The repulsion between negatively charged SC2RBD-2 and metamaterial disrupts effective molecular detection. The lowest detection limit for all receptor binding domains is 0.1 mg/mL (41.7 μM).

Experimental results show that within a few minutes, variants with similar structures can be identified by detecting the terahertz spectral changes of the SARS-CoV-2 and SARS-CoV spike protein receptor binding domains. The development of a COVID-19 virus diagnostic technique using terahertz time-domain spectroscopy to detect infectious diseases and track variant strains is possible.

From the above analysis, terahertz time-domain spectroscopy has great potential in qualitative and quantitative identification of key features in the biomedical field.

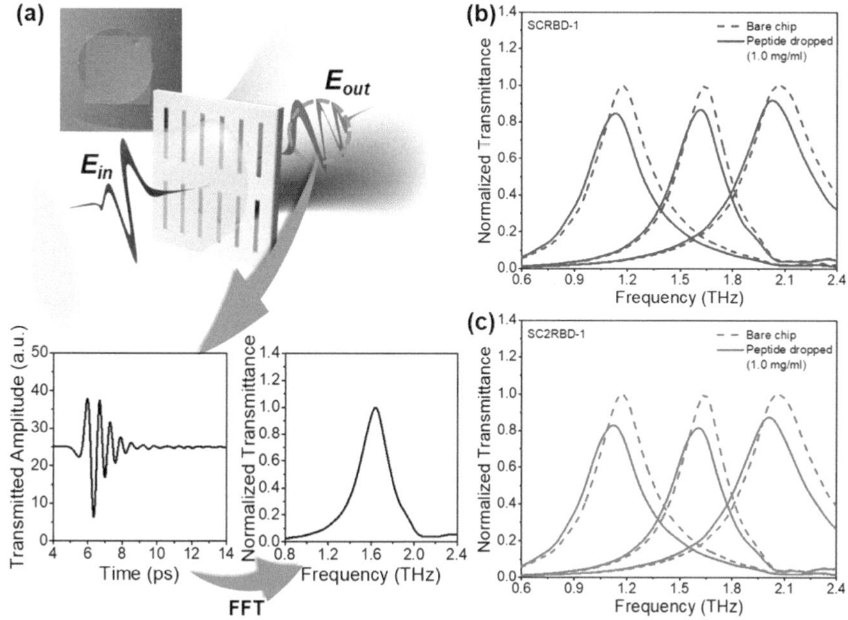

Fig. 6.29 **a** Sensing process of THz-TDS in transmission mode **b** Terahertz spectra of SCRBD-1 and **c** SC2RBD-1 at 1.0 mg/mL at 1.16 THz, 1.64 THz, and 2.07 THz. Reprinted from Ref. [22], ©2022, under CC-BY 4.0

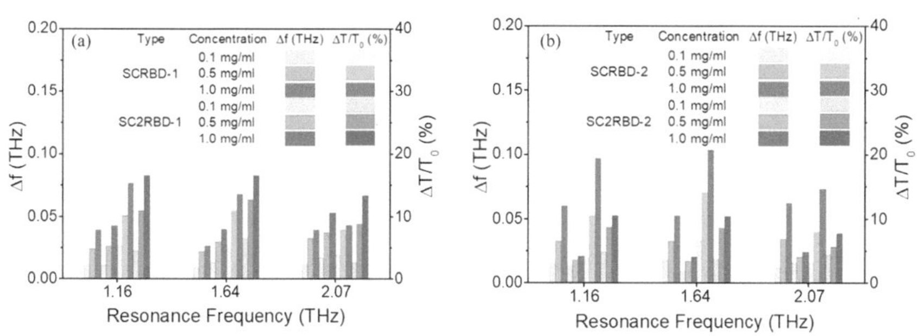

Fig. 6.30 **a** Terahertz spectral shifts and relative transmittance changes of SCRBD-1 and SC2RBD-1 **b** Terahertz spectral shifts and relative transmittance changes of SCRBD-2 and SC2RBD-2. Reprinted from Ref. [22], ©2022, under CC-BY 4.0

However, due to the numerous components in biological samples and interference from moisture, the signal-to-noise ratio of the target substance's absorption peak is too small to be recognized, and it has not yet truly entered clinical applications. There is a need to further improve the sensitivity and dynamic range of detection.

To improve the sensitivity of terahertz time-domain spectroscopy detection, efforts can be made in sample processing, data analysis, and system optimization [23].

Sample processing includes attenuated total reflection spectroscopy, metamaterial sensing, and the addition of reagents. Attenuated total reflection spectroscopy is sensitive to small amounts of samples, but cannot specifically enhance the resonance of the target substance. Metamaterials with fixed resonant frequencies can reflect changes in the total dielectric constant and can only be used for quantitative analysis, while metamaterials with adjustable resonant frequencies can be used to obtain the spectrum of the sample for qualitative and quantitative analysis, but the spectral frequency range is limited by the material. Reagents can enhance the signals of different substances but may introduce sample contamination. In data analysis, various noise reduction and reconstruction algorithms are primarily employed to extract meaningful information from spectral data and improve the signal-to-noise ratio. Algorithms can extract valuable information from spectral data; however, some useful details may be lost in the process. In system optimization, improving the energy output of the terahertz source or enhancing the sensitivity of the terahertz detector is the primary approach to directly increase the signal-to-noise ratio during measurement. This is the most direct method to enhance the terahertz signal, but the overall enhancement still causes the signal of the target substance to be drowned out by the signals of other substances.

References

1. Yang K, Li JN, Chapelle MLDL, et al. A terahertz metamaterial biosensor for sensitive detection of microRNAs based on gold-nanoparticles and strand displacement amplification. Biosens Bioelectron. 2021;175: 112874.
2. Pan TT, Li SP, Zou T, et al. Terahertz spectra of l-phenylalanine and its monohydrate. Spectrochim Acta Part A Mol Biomol Spectrosc. 2017;178:1386–425.
3. Zhu ZJ, Ren GH, Cheng C, et al. Broadband terahertz spectroscopy study of L-dopa. Chinese Laser. 2019;46(6):0614017.
4. Liu R, He MX, Su RX, et al. Insulin amyloid fibrillation studied by terahertz spectroscopy and other biophysical methods. Biochem Biophys Res Commun. 2010;391(1):862–7.
5. Zhu ZJ, Cheng C, Chang C, et al. Characteristic fingerprint spectrum of neurotransmitter norepinephrine with broadband terahertz time-domain spectroscopy. Analyst. 2019;144(8):2504–10.
6. Wang PF, Wang XC, Liu LY, et al. The hydration shell of monomeric and dimeric insulin studied by terahertz time-domain spectroscopy. Biophys J. 2019;117(3):533–41.
7. Zhou RY, Wang C, Huang YX, et al. Label-free terahertz microfluidic biosensor for sensitive DNA detection using graphene-metasurface hybrid structures. Biosens Bioelectron. 2021;188(113336):0956–5663.
8. Reid CB, Reese G, Gibson AP, et al. Terahertz time-domain spectroscopy of human blood. IEEE J Biomed Health Inform. 2013;17(4):774–8.
9. Jeong K, Huh YM, Kim SH, et al. Characterization of blood using terahertz waves. J Biomed Opt. 2013;18(10): 107008.
10. Lian Z, Niu H, Li S. Detection of ageing effect in polyesterimide by terahertz time-domain spectroscopy. IEEE Trans Terahertz Sci Technol. 2019;9(6):651–8.
11. Shi W, Wang Y, Hou L, et al. Detection of living cervical cancer cells by transient terahertz spectroscopy. J Biophotonics. 2021;14(1): e202000237.
12. Shiraga K, Suzuki T, Kondo N, et al. Hydration state inside HeLa cell monolayer investigated with terahertz spectroscopy. Appl Phys Lett. 2015;106: 253701.

13. Qin JY, Xie LJ, Ying YB. Rapid analysis of tetracycline hydrochloride solution by attenuated total reflection terahertz time-domain spectroscopy. Food Chem. 2017;224:262–9.
14. Cao Y, Chen J, Zhang G, et al. Characterization and discrimination of human colorectal cancer cells using terahertz spectroscopy. Spectrochim Acta Part A Mol Biomol Spectrosc. 2021;256: 119713.
15. Zou Y, Liu Q, Yang X, et al. Label-free monitoring of cell death induced by oxidative stress in living human cells using terahertz ATR spectroscopy. Biomed Opt Express. 2018;9(1):14–24.
16. Zhang C, Liang L, Ding L, et al. Label-free measurements on cell apoptosis using a terahertz metamaterial-based biosensor. Appl Phys Lett. 2016;108(24): 241105.
17. Yan X, Yang M, Zhang Z, et al. The terahertz electromagnetically induced transparency-like metamaterials for sensitive biosensors in the detection of cancer cells. Biosens Bioelectron. 2019;126:485–92.
18. Zhang J, Mu N, Liu L, et al. Highly sensitive detection of malignant glioma cells using metamaterial-inspired THz biosensor based on electromagnetically induced transparency. Biosens Bioelectron. 2021;185: 113241.
19. Wang Y, Sun Z, Xu D, et al. Detection of cerebral ischemia based on terahertz time-domain spectroscopy system. Acta Optica Sinica. 2020;40(4):0430001.
20. Zhao HL, Wang YY, Chen LY, et al. High-sensitivity terahertz imaging of traumatic brain injury in a rat model. J Biomed Opt. 2018;23(3): 036015.
21. Zhou ZT, Zhou T, Zhang S, et al. Multispectral Imaging: multicolor T-ray imaging using multispectral metamaterials. Adv Sci. 2018;5(7):1870044.
22. Lee SH, Lee YK, Lee S-H, et al. Detection and discrimination of SARS-CoV-2 spike protein-derived peptides using THz metamaterials. Biosens Bioelectron. 2022;202: 113981.
23. Peng Y, Shi C, Zhu Y, et al. Terahertz spectroscopy in biomedical field: a review on signal-to-noise ratio improvement. PhotoniX. 2020;1:12.
24. Yang JQ, Li SX, Zhao HW, et al. Molecular recognition and interaction between uracil and urea in solid-state studied by terahertz time-domain spectroscopy. J Phys Chem A. 2014;118(46):10927–33.

Chapter 7
Application of THz-TDS in the Field of Cultural Heritage Preservation

The preservation and inheritance of cultural relics cannot be separated from the usage of technology. Non-destructive testing techniques can be used to measure and analyze cultural relics, providing information about their physical structure, chemical composition, and historical age. Commonly used methods for non-destructive testing of cultural relics include ultrasonic testing, X-ray fluorescence spectrometry (XRF), Raman spectroscopy, and Fourier transform infrared spectroscopy (FTIR). Ultrasonic technology is suitable for detecting the internal structure of stone materials. XRF can penetrate porcelain, calligraphy and painting, and bronze ware to detect mineral pigment, but it cannot distinguish wall painting pigments containing the same elements. Raman spectroscopy can also analyze pigment components, but it is easily interfered by fluorescence when dealing with organic pigments. Fourier transform infrared spectroscopy can distinguish pigments with the same elements, but the mid-infrared region of infrared spectroscopy rarely shows characteristic peaks of inorganic pigments, such as oxides and sulfides. The above non-destructive testing techniques are difficult to penetrate the surface of cultural relics to reach the interior. Most murals are multi-layered structures. In addition to the surface pigments, the parts below the surface are also very important for analysis. THz-TDS can penetrate murals and analyze mural pigment components, etc [1, 2].

This chapter introduces the application of THz-TDS technology in cultural heritage preservation, such as murals painting, pottery glaze removal, painting restoration, and hidden inscription detection on crosses. Cultural relics such as murals, ceramics, and woodblock prints are all non-polar materials. Terahertz waves can penetrate non-polar materials, making them important in cultural heritage preservation. Moreover, terahertz wave is none-ionizing radiation, so it will not damage the samples being tested. It can be used as a substitute for X-rays in the detection and research of fragile cultural relics such as ceramics, paintings, and books. Specifically, terahertz technology can be applied for imaging of the material composition, hidden layers, and internal structure of murals, detecting internal defects and cracks in ceramics, and real-time monitoring of the condition of paintings [3]. Compared

T. Chen et al., *Terahertz Time Domain Spectroscopy*,
https://doi.org/10.1007/978-981-95-2605-5_7

with infrared imaging, terahertz imaging has better penetration ability and can image the hidden layers of cultural relics. Recently, compressed sensing imaging algorithms have been able to effectively achieve rapid imaging. Using algorithm enhancement and other means can help further increase the accuracy of THz-TDS imaging in detecting cultural relics.

7.1 Mural Peeling Detection

The earliest application of terahertz technology in mural detection was in the Takamatsuzuka mural detection in Japan. In 1972, the Takamatsuzuka tomb was discovered in Nara, Japan. According to excavation and research results, Takamatsuzuka is a two-layer circular tomb with diameters of 23 m (lower layer) and 18 m (upper layer). The tomb contains a stone chamber made of sixteen tuff blocks. The walls and ceiling of the stone chamber are all plastered with gypsum; the depth of the gypsum layer is usually a few millimeters. The gypsum is painted with extremely colorful murals, the content of the murals inside the stone chamber symbolizes the ancient Chinese mythical beasts and four gods that rule the four directions, with the east wall being the Azure Dragon, the west wall being the White Tiger, the north wall being the Black Tortoise, and the east and west walls are painted with groups of women and men, and the ceiling is also painted with constellations. The Takamatsuzuka murals attracted much attention, and in 1974 Takamatsuzuka was designated as a national treasure of Japan. Figure 7.1a is a photo of a group of lady murals painted on the west wall; this photo was taken shortly after the murals were discovered. After the murals were excavated, Japan established a special committee to discuss how to protect this mural in the future. It was finally decided to preserve the mural on site. In 1976, protective facilities were built in front of the stone chamber to protect the murals from fluctuations in temperature and humidity.

After the soil near the stone chamber was reinforced in 2001, black mold was found both inside and outside the stone chamber. In December 2001, black mold was also found on the murals. The composition of the mold includes genera such as Fusarium, Trichoderma, and Penicillium. The main reason for the growth of mold is believed to be the increase in temperature in the stone chamber caused by soil reinforcement. The growth of mold is an urgent problem. To protect the murals, the committee decided to dismantle the stone chamber to repair the murals. The murals were dismantled and transported to a new laboratory in the state-run Asuka Historical Park near the Takamatsuzuka tomb. Figure 7.1b was taken before the murals were moved, and mold was found all over the pattern of the lady in the mural.

Before the mural repair and cleaning, it is necessary to understand the condition of the murals and consider how to control the environment to protect the murals. Starting from 2007, various scientific methods have been applied to mural condition detection, such as microscope observation, digital scanning of visible images to record the current condition of the murals, and X-ray fluorescence spectrometers for element analysis.

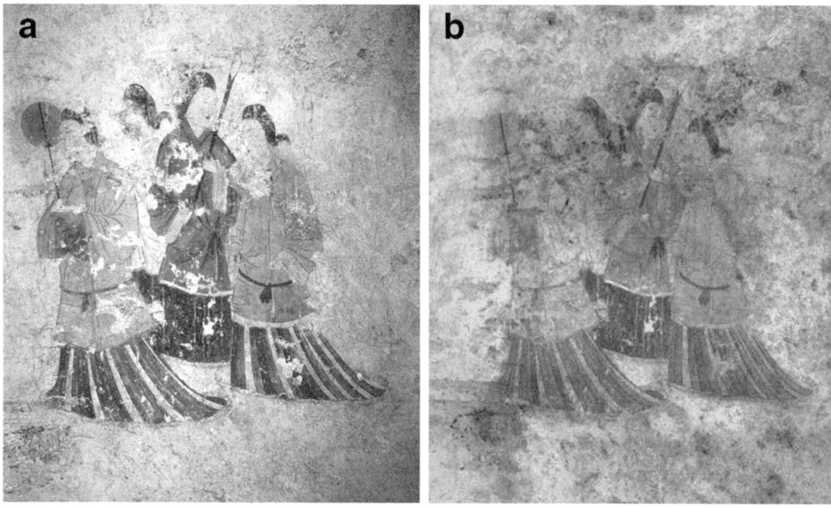

Fig. 7.1 Mural of the Lady from the Takamatsuzuka Tomb **a** Photographed in 1972 when first excavated, **b** Photographed in 2001 before disassembly. Reprinted with permission from Springer Nature from Ref. [4], ©2023. Permission conveyed via RightsLink, licence no. 502005491, 31 July 2025

When the mural was in the stone chamber, the relative humidity was almost 100%. One of the important components of the mural is the gypsum layer. After the mural was moved, the relative humidity around the mural decreased, and the condition of the gypsum layer may change. For example, its strength may decrease due to a reduction in water content. The gypsum layer may also peel off due to shrinkage. Even if the surface of the gypsum looks hard, peeling can occur inside. Initially, restorers identified peeling locations by touch, which is highly subjective. It is particularly important to objectively and independently check the peeling locations through scientific means, which can serve as a cross-supplement to the objective judgment of restorers. Terahertz waves can penetrate the gypsum layer, and terahertz imaging is a suitable method for non-destructive inspection of the condition and peeling location of the gypsum layer. The Azure Dragon mural on the east side of the stone chamber was detected by terahertz technology [4]. The cut Azure Dragon mural was fixed face up, and the terahertz scanning head, mechanical stage, and system controller were installed on a frame specially made for studying the Takamatsuzuka murals, as shown in Fig. 7.2.

During the terahertz spectroscopic imaging measurement process, the terahertz scanning head moves above the mural. The size of the Azure Dragon pattern mural is 1160 mm × 910 mm, and the terahertz scanning head can move within a 250 mm × 250 mm area on the mechanical carrier stage moved within an area of mm. Therefore, the area of the mural pattern to be tested was divided into 4 × 5 parts, and a total of 20 imaging measurements were carried out. The results contain spectral and imaging

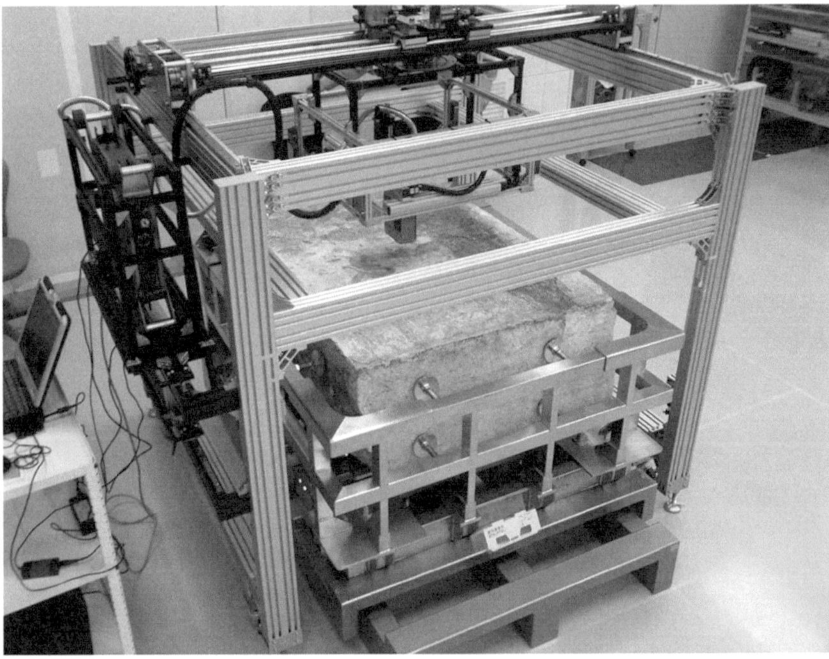

Fig. 7.2 Detection of cut murals using a THz-TDS system and a custom mobile probe. Reprinted with permission from Springer Nature from Ref. [4], ©2023. Permission conveyed via RightsLink, licence no. 502005491, 31 July 2025

information, and the imaging results are shown in Fig. 7.3. In the terahertz imaging results, the solid clay layer and tuff can be seen.

To further understand the condition and peeling situation of the Azure Dragon mural, a point-by-point analysis of the terahertz time-domain spectral imaging results of the Azure Dragon mural was carried out. The Azure Dragon mural in Fig. 7.4 is a representative result. Point A is a solid clay layer, the clay layer at point B has been lost, and the surface of the tuff can be seen; at point C, the tuff cannot be seen, and it looks very solid from the surface.

Analyzing the Terahertz time-domain waveforms of the three points on the left side of Fig. 7.5, the terahertz pulse peak of point A appears around 25 ps, which is the reflection peak of the surface of the solid clay layer; the terahertz pulse peak of point B is at 55 ps, which is the reflection peak of the tuff surface. The terahertz pulse signal at point C shows that in addition to the terahertz reflection peak of the solid clay layer, there is also a reflection peak around 47 ps. It indicates that the clay layer has peeled off from the tuff. Comparing the cross-sectional imaging corresponding to points A, B, and C on the right side of Fig. 7.5, it can be found that the boundary between the clay layer and the tuff at point C is between the surface of the clay layer and the tuff. Furthermore, it verifies that the clay layer at point C has peeled off from

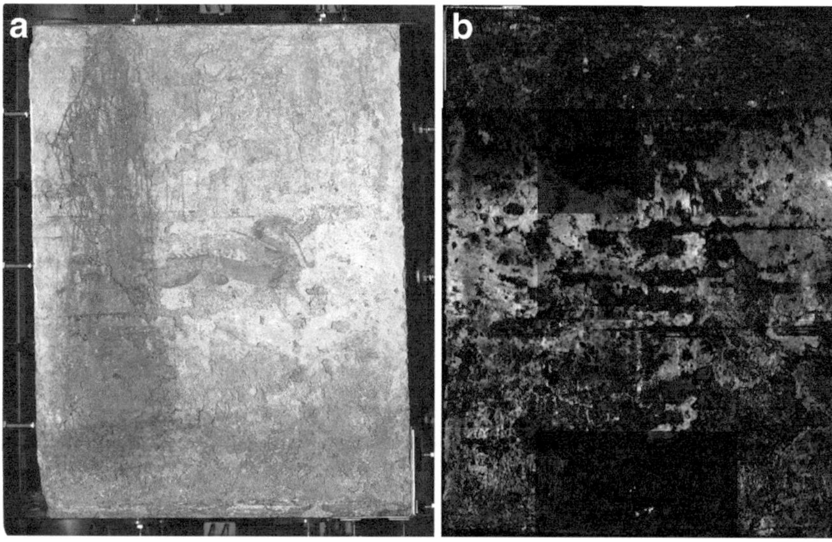

Fig. 7.3 Terahertz detection results of the Azure Dragon mural. **a** Physical image of the mural **b** Terahertz imaging results. Reprinted with permission from Springer Nature from Ref. [4], ©2023. Permission conveyed via RightsLink, licence no. 502005491, 31 July 2025

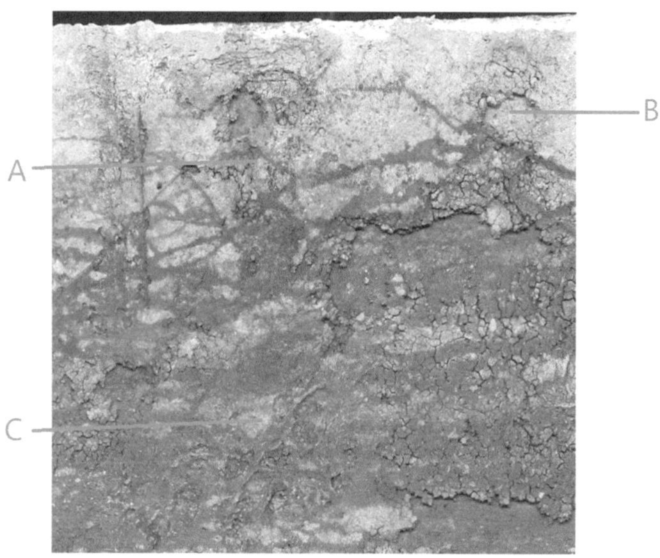

Fig. 7.4 Different peeling states at various locations of the Azure Dragon mural. Reprinted with permission from Springer Nature from Ref. [4], ©2023. Permission conveyed via RightsLink, licence no. 502005491, 31 July 2025

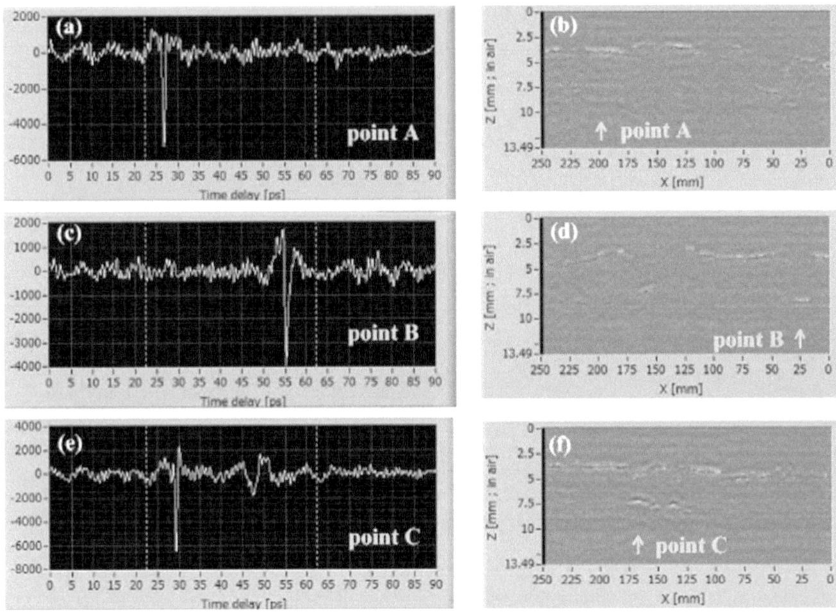

Fig. 7.5 Terahertz time-domain pulse response and cross-sectional imaging at different locations of the Azure Dragon mural. Reprinted with permission from Springer Nature from Ref. [4], ©2023. Permission conveyed via RightsLink, licence no. 502005491, 31 July 2025

the tuff. It is the unique penetration of terahertz technology that allows us to see the peeling state inside the clay layer.

7.2 Mural Pigment Detection in Dazhao Temple

In addition to the application of terahertz technology for mural peeling detection in the Takamatsuzuka in Japan, Chinese researchers have also done a lot of work in mural pigment detection, such as the Dazhao Temple in Inner Mongolia, Dunhuang Grottoes mural detection, etc. In the murals of Dazhao Temple in Inner Mongolia, THz-TDS technology is applied to the detection of mural pigment types, which plays an important role in the age identification of murals [5].

Dazhao Temple is in Dazhao Qian Street, Yuquan District, Hohhot City, Inner Mongolia Autonomous Region. It is a Gelugpa temple of Tibetan Buddhism. Dazhao is the first of the "Seven Temples" in the Hohhot area during the Ming and Qing Dynasties. The temple palace is surrounded by brilliant murals. The identification of mural pigments is of great significance to the composition and age of murals. Figure 7.6 is a physical picture of the mural fragments of Dazhao Temple. Through past analysis, it is known that a~e contain different mural pigments, among which a is Pb_3O_4 component minium pigment, b is HgS component cinnabar red, c is malachite, d is orpiment, and e is calcite. Terahertz spectral analysis of these five types of pigment samples, as shown in Fig. 7.7, reveals that a minium pigment and b cinnabar red pigment contain metal components, and their reflectivity to terahertz is higher than that of malachite c, orpiment d, and calcite e.

In addition to the known a~e pigments, the component analysis of other parts still needs to be explored. To further distinguish the pigment composition of other areas, terahertz spectral imaging of mural samples was carried out, and the imaging results are shown in Fig. 7.8.

From the terahertz imaging results of the mural fragments in Fig. 7.8, there are clay blocks in the upper left corner of the mural sample. The surface of the clay is

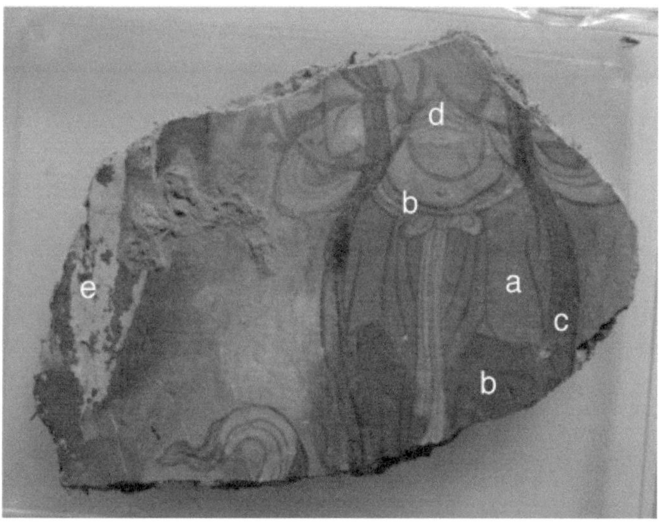

Fig. 7.6 Physical image of mural fragments from the Dazhao Temple, **a–e** include different pigments. Reprinted/Adapted from Ref. [5], ©2010, under CC-BY 4.0

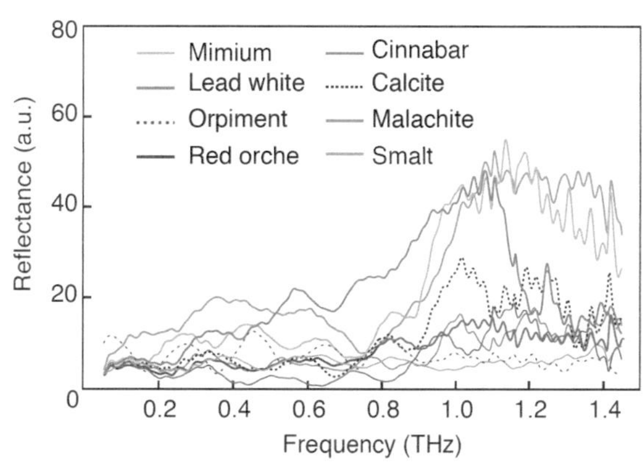

Fig. 7.7 Terahertz reflection spectra of different pigments. Reprinted/Adapted from Ref. [5], ©2010, under CC-BY 4.0

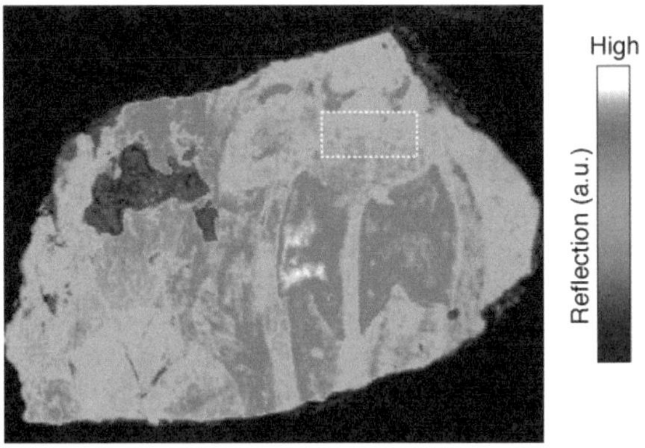

Fig. 7.8 Terahertz imaging results of the mural. Reprinted/Adapted from Ref. [5], ©2010, under CC-BY 4.0

uneven and causes a diffuse reflection of the terahertz signal. The terahertz spectral imaging system cannot effectively receive the terahertz signal, and the reflectivity in the imaging results is very low. Comparing the physical picture of the mural and terahertz imaging, it can be found that the yellow belt of the Buddha image in the mural does not show in the terahertz imaging, which is outlined in the figure with a white dashed box. It can be judged that the belt pigment is different from the pigment in the minium area a, and it should be an organic dye.

To further measure the accuracy of this judgment, an X-ray fluorescence spectrometer (XRF), commonly used for element analysis, was used to analyze the composition of the mural. Figure 7.9 is the result of XRF element imaging. From the XRF element imaging results of the mural, the lead element in the minium area a is not found in the belt area, which is consistent with the result of terahertz spectral imaging. Terahertz imaging can be applied to the element analysis of murals, and it has important guiding significance for the age, status, and component analysis of murals.

7.3 Pottery Glaze Peeling Detection

Since the Middle Ages, monochrome glazed pottery art has been produced in northern Germany and widely used as architectural elements locally. In practical use, the glaze is prone to falling off and damage. Studies have shown that salt and gypsum crystallization are the main causes of glaze damage. As salinity increases, it leads to the loss of terracotta surface material caused by calcium and sulfate ions, and these salt corrosives mainly exist in the glazed area. Among other influencing factors such as weather changes and air pollutants, salt crystallization is the reason for the detachment of the glaze part. It is necessary to fully understand the adhesion/separation between terracotta and glaze to decide whether the restoration method needs to be backfilled to prevent the loss of glaze. As a typical application case, the

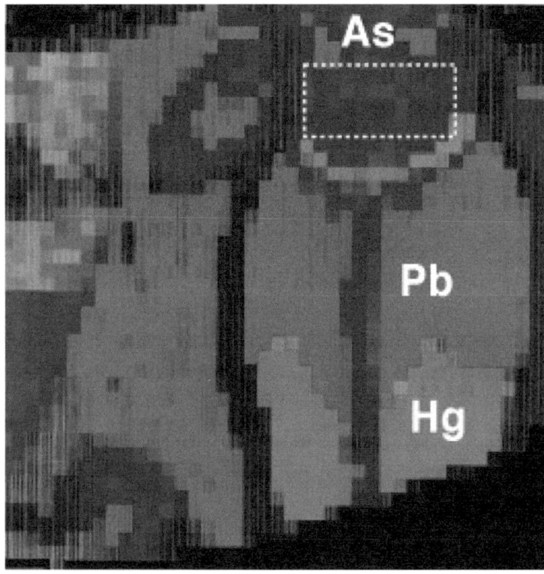

Fig. 7.9 X-ray fluorescence spectrometer imaging results of the mural (As, Pb, Hg element analysis). Reprinted/Adapted from Ref. [5], ©2010, under CC-BY 4.0

glazed medal decorating the exterior wall of a row house in Lower Saxony, Germany, was examined for voids and deglazing using terahertz technology.

A reflective THz-TDS imaging system was used for non-destructive analysis of the glazed medal, the medal is a medieval character portrait. The shoulder area was selected for terahertz non-destructive analysis, and each row of the dashed box A~I was scanned for imaging, yielding terahertz spectroscopy imaging results [6].

When pulsed terahertz irradiates the glazed pottery, part of the terahertz signal is reflected on the glaze surface and received by the THz-TDS system, while another terahertz pulse penetrates the glaze and continues to propagate. If there is a deglazing air layer, the terahertz model continues to propagate to the pottery surface, and most of the signal is reflected and detected by the THz-TDS system. By analyzing the time difference Δt of the terahertz pulse peak signals on the glaze surface and the pottery surface, it is possible to determine whether there are voids and confirm the spacing of the voids. If the terahertz pulse signal cannot distinguish the reflection peaks of the glaze surface and the pottery surface, then the corresponding medal area is tightly fitted between the glaze and the pottery and there are no voids. If the interval between the terahertz pulse signals on the glaze surface and the pottery surface is very long, then there are voids and deglazing in the corresponding medal area, and the time difference between the two pulse signals corresponds to the spacing of the deglazing voids. Point-by-point scanning of each row of shoulder area A~I of the glazed medal yielded terahertz imaging of the corresponding glaze and pottery glaze cross-sections. The imaging results show that there are deglazing phenomena in the dashed areas a~i, and the contours of the deglazing areas can be depicted by analyzing the imaging results.

A specific area of the glazed medal sample was selected for analysis to further analyze the width of the deglazing area. The refractive index n of the glaze was measured in advance, and the time difference Δt of the terahertz pulse peak signals on the glaze surface and the pottery surface was converted into the spacing of the deglazing area, eliminating the transmission time delay on the glaze surface, corresponding to a deglazing area width of 1.74 mm. The same area of the glazed medal was scanned with X-ray tomography, yielding a deglazing area width of 1.8 mm, which is consistent with the results of THz-TDS imaging.

Compared with X-ray tomography, THz-TDS imaging has low photon energy and no ionizing radiation, so it will not damage the sample and is safe for the operator. In addition, X-ray tomography scanners are generally large and not portable, making it impossible to measure glazed pottery inside walls on-site, while THz-TDS imaging systems are relatively flexible and can be carried to the site for measurement.

7.4 Restoration and Detection of Painted Artifacts

Terahertz imaging can not only obtain two-dimensional information but also determine the thickness of the repainted pigment layer by analyzing the flight time information contained in each waveform, and separate each layer in chronological order. The discovery of this hidden information is crucial for archaeology and cultural heritage. With the inscription records, archaeologists can speculate on the painting restoration time and era, further understand the story behind the painting, and more realistically restore the painting creation process. Terahertz imaging was used to scan and analyze the painting works of the 18th-century neoclassical painter Nicolas Abidhao, revealing the repainted area [7]. Terahertz technology combined with near-infrared imaging also revealed the repainted area when testing Picasso's oil painting "Homme au chapeau", and terahertz time-domain imaging can provide more details of the draft layer and painting preparation layer [8]. Terahertz technology was used to test four easel paintings in a Mexican church, revealing that these paintings have been modified twice since their completion in 1725, and a hidden record inscription was also found in the painting [9].

Terahertz imaging technology can also be applied to the post-restoration assessment of artifacts such as paintings. Many art restorations can only be technically evaluated by technicians with extensive professional knowledge and experience. Terahertz waves can penetrate painted artifacts and serve as a diagnostic method for evaluating the structural layers of paintings, aiding in the judgment and management of the restoration of paintings and other artifacts.

As a typical case, we introduce the use of terahertz technology in the restoration process detection of a painting of a female scholar collected in the Bordeaux Museum in France. The authors report that the oil painting—depicting a young female scholar—shows no discernible cracks, pigment loss, or other visible defects on its surface. On the back of the painting, there is a text description "Hortense Mancini, niece du Cardinal Mazarin par Pierre Mignard", indicating that the work is from the

18th century and depicts the niece of Cardinal Mazarin, the prime minister during the reign of King Louis XIV of France. The back of the painting uses a wooden frame, and terahertz waves cannot penetrate wood. The black part is the shape of the frame behind the painting, and the irregular black shadow on the left corresponds to the reinforcement parts behind the frame. The droplet shaped irregular pattern represents the adhesive used in past renovations of the painting, such as rabbit skin glue.

To further assess the painting's condition, the team employed a reflective terahertz imaging system, which highlights defects with greater clarity. Six solid red circles pinpoint areas of concern: the upper-left and lower-right circles reveal canvas gaps that extend into the ground layer and lining fabric, while the upper-right circle, drawn with an L-shaped dashed line, marks canvas cracking that affects both paint and primer layers. After restoration, a follow-up terahertz scan—shown on the right—repeats these positions with red dashed circles; the previously darkened shadows have almost vanished, confirming that the conservation effort was successful. The project subsequently earned coverage by the BBC and other outlets, reinforcing its international significance.

7.5 Hidden Inscription Detection on Cross

The corrosion layer on lead artifacts and archaeological items can seriously hinder the interpretation of inscriptions, thereby obstructing a comprehensive understanding of the items and their context. The oxidation of lead that causes corrosion may be reduced through reversible chemical reactions. However, corrosion damage is generally not completely reversible, and the reduction process is not easy or reversible in all cases. In early 2022, researchers from France and the United States used the unique penetration ability and broadband coverage of terahertz pulses for non-destructive terahertz multispectral imaging to observe the corrosion of a 16th-century funeral cross from Lorraine, France. Multispectral images obtained from different terahertz bands were input into a carefully designed post-processing chain for image restoration and enhancement, thereby achieving the first reading of potentially lost blurred inscriptions. The combination of terahertz characteristics of the constituent materials and advanced signal image processing technology opens a new perspective for multi-resolution analysis at terahertz frequencies. As a technology in archaeological research, it will ultimately provide unprecedented information for digital acquisition and recording, character extraction, classification, and recognition.

By utilizing the broadband characteristics of terahertz pulses, multispectral images can be obtained from different frequency bands, and surface and subsurface lead features can be characterized with different spatial resolutions. Features related to the inscription can only be recognized at high frequencies, and additional information provided at low frequencies can be input into the image post-processing chain to improve the visibility of the inscription. Terahertz time-domain spectroscopic imaging can be applied to various cultural heritages, providing valuable materials for archaeological research, and can also be used for documentation, preservation, and restoration.

The size of the cross is approximately 115 mm × 115 mm, with a thickness of about 2 mm. The white appearance is due to corrosion. The horizontal arm of the cross was chosen as the area of interest for terahertz imaging, as shown in Fig. 7.10a. Using reflective measurements, terahertz pulse imaging emits single-cycle terahertz pulses to detect surface and subsurface information of the lead cross. Figure 7.10b provides the system's reference terahertz pulse, measured by placing a metal plate at the sample position. A raster scan was performed with a set of 0.2 mm steps moving in the X and Y directions. After the raster scan, a three-dimensional volume of reference data was obtained. The irregular changes in the terahertz reflection spectrum correspond to the discontinuities and roughness at the air-corrosion layer and corrosion-lead interfaces contain information related to the inscription. The power spectrum of the terahertz pulse ranges from about 60 GHz to 3.0 THz, as shown in Fig. 7.10c. The full spectrum of the broadband terahertz pulse is divided into six frequency bands, each with a bandwidth of approximately 0.5 THz. The terahertz multi-spectral image is obtained based on the frequency components of each band. The size of the terahertz light spot is related to the frequency, and at 1 THz, its full width at half maximum (FWHM) is about 2.42 mm.

The electrochemical method is used to reduce the corrosion of the lead cross sample. The optical images obtained before and after the reverse corrosion are compared to verify the terahertz imaging results, as shown in Fig. 7.11. The image

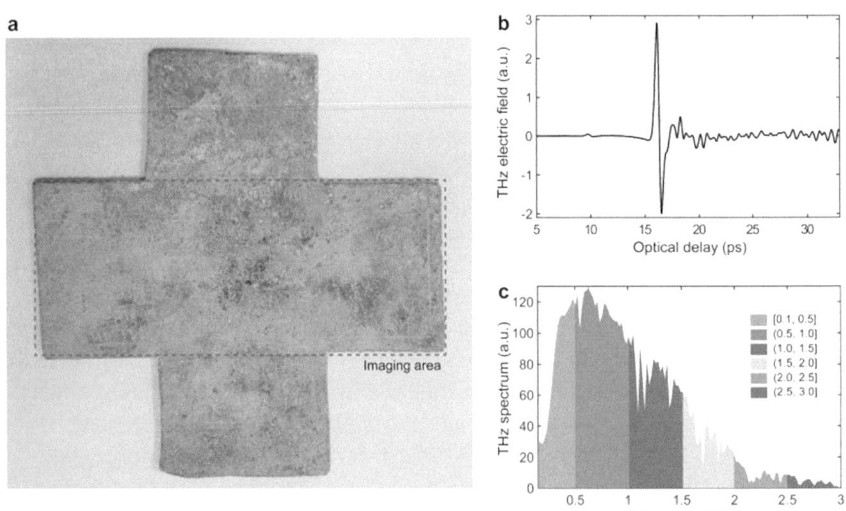

Fig. 7.10 **a** Photograph of the lead cross sample used in the study **b** Terahertz pulse reference signal used in the system **c** Power spectrum of the reference terahertz pulse. Reprinted/Adapted from Ref. [10], ©2022, under CC-BY 4.0

obtained after post-processing can reveal most of the features related to the inscription. The terahertz image shown in Fig. 7.11b can reveal almost all the features, and the elimination of corrosion can be observed, as shown in Fig. 7.11c. As shown in the area highlighted by the red dashed frame in Fig. 7.11a, the terahertz image successfully displays some inscriptions that are almost invisible on the original lead cross. Based on the observed features, the inscription is carved in Carolingian minuscule script, in Latin. From the Charles de Bruyère Museum, the Latin words "tuum fat voluntas tua" and part of "quotidianum" have been identified, based on the terahertz features in the red rectangle in the upper right, and part of "dimittimus" and "tentationem", based on the terahertz features in the bottom frame. All these texts come from the Lord's Prayer, and after the corrosion products of the cross were removed, the correctness of these texts was confirmed by inspection. Experts call these plaques identity cards because many of their inscriptions mention the deceased. The research results show that terahertz imaging can visualize blurred inscriptions on corroded lead crosses in a non-destructive and non-contact manner. It is worth noting that the successful revelation of blurred inscriptions depends on the extraction of high-frequency components of broadband terahertz pulses. At the same time, even if the low-frequency bands lack the resolution to fully capture the details of the inscription itself, these bands also need to remove confusing irrelevant features from the image.

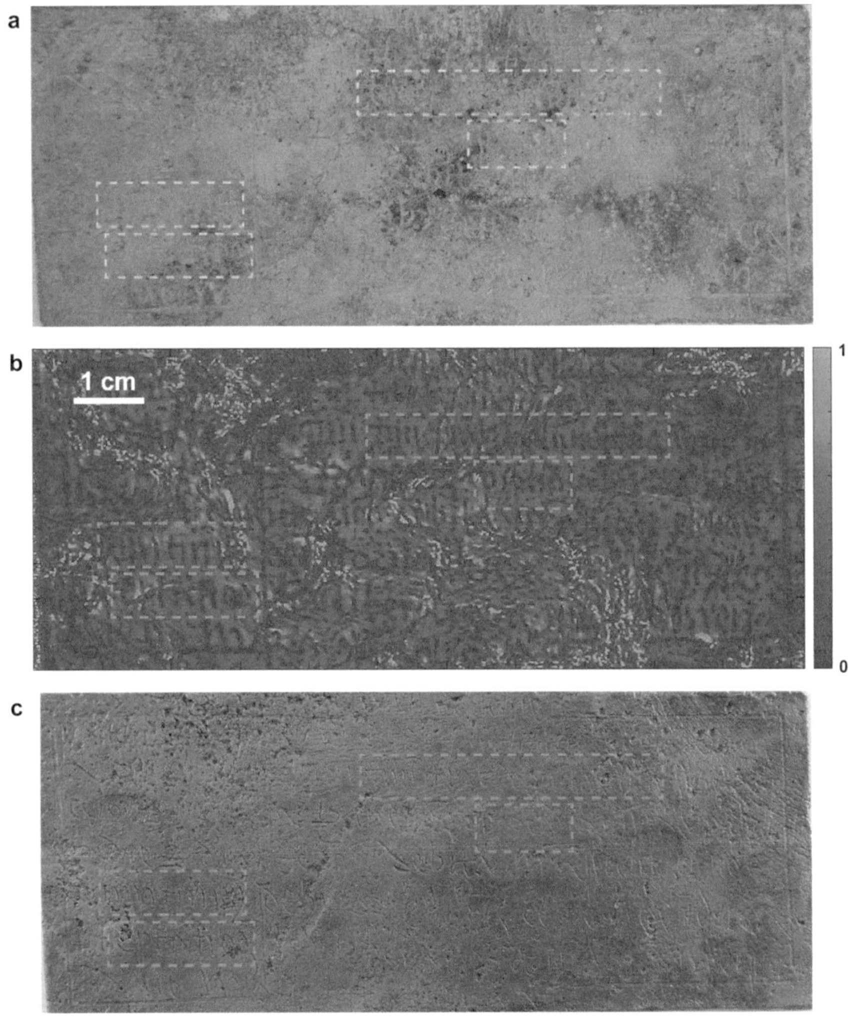

Fig. 7.11 Comparison between optical and terahertz images of the visualized inscription on the cross. **a** Photograph of the original lead cross **b** Terahertz imaging of the processed lead cross **c** Photograph of the lead cross after chemical corrosion removal. Reprinted/Adapted from Ref. [10], ©2022, under CC-BY 4.0

References

1. Yang J, Qiu J, Jiang H, Hu M, Liu S, Zhang H. Application and prospect of terahertz technology in paintings. J Infrared Millim Waves.2022;41:218–229.
2. Krugener K, Ornik J, Schneider LM, et al. Terahertz inspection of buildings and architectural art. Appl Sci. 2020;10:5166.
3. Zhou P, Qi Y, Li ZY, et al. Progress of terahertz technology in the field of cultural heritage. Cult Relics Prot Archaeol Sci. 2016;28:133–143.

4. Inuzuka M, Kouzuma Y, Sugioka N, et al. Investigation of layer structure of the Takamatsuzuka mural paintings by terahertz imaging technique. J Infrared Millim Terahertz Waves. 2017;38:380–9.
5. Fukunaga K, Hosako I, Kohdzuma Y, et al. Terahertz analysis of an East Asian historical mural painting. J Eur Opt Soc Rapid Publ. 2010;5:10024.
6. Kirsti K, et al. Non-destructive analysis of material detachments from polychromatically glazed terracotta art-work by THz time-of-flight spectroscopy. J Infrared Millim Terahertz Waves. 2017;38:495–502.
7. Dandolo CLK, Filtenborg T, Fukunaga K, et al. Reflection terahertz time-domain imaging for analysis of an 18th century neoclassical easel painting. Appl Opt. 2015;54(16):5123–9.
8. Kaori F, et al. THz pulsed time-domain imaging of an oil canvas painting: a case study of a painting by Pablo Picasso. Appl Phys A. 2016;122:106.
9. Sepulveda AM, et al. History of Mexican easel paintings from an altarpiece revealed by non-invasive terahertz time-domain imaging. J Infrared Millim Terahertz Waves. 2017;38:403–12.
10. Dong J, Ribeiro A, Vacheret A, et al. Revealing inscriptions obscured by time on an early-modern lead funerary cross using terahertz multispectral imaging. Sci Rep. 2022;12(1):3429.

Chapter 8
Application of THz-TDS Technology in Nondestructive Testing Field

Nondestructive testing (NDT) is a technology that uses instruments based on principles such as radiation, ultrasound, infrared, electromagnetic, and Terahertz nondestructive testing to inspect materials, components, and equipment for defects, chemical, and physical parameters without affecting their future functionality or current operational status. NDT is an indispensable tool for industrial development. It reflects a country's industrial development level, and its importance is internationally recognized. The main types of NDT include radiographic inspection, ultrasonic testing, magnetic particle inspection, and liquid penetrant inspection. Other commonly used NDT methods include eddy current testing, acoustic emission testing, thermographic infrared testing, and leak testing. With the emergence of various new materials and technologies, traditional NDT methods are increasingly unable to meet the testing needs of new technologies, and some new NDT methods have begun to emerge. Terahertz waves have excellent penetration for non-polar materials and can penetrate composite and non-metallic materials such as glass, cardboard, carbon fiber, car paint, and fabric. THz-TDS non-destructive testing is widely used in many industries and has a very broad development prospect and space.

Terahertz non-destructive testing is a technique that uses terahertz electromagnetic radiation equipment for inspection, analysis, and evaluation. THz-TDS imaging technology can be used to assess the properties of materials, components, and systems without causing damage to them. In industries such as papermaking, aerospace, automotive and broadcasting, pharmaceuticals, semiconductors, security, and electrical power, THz-TDS technology is used for non-destructive analysis and quality control of non-polar materials. THz-TDS technology can effectively inspect the delamination of surface pigments and inner cores, detect defects in ceramics and composite materials, and scan the physical structure of insulating materials. Non-destructive evaluation using THz-TDS technology can inspect multi-layer structures and identify foreign bodies, inclusions, delamination, mechanical impact damage, thermal damage abnormalities, and liquid intrusion. As shown in Fig. 8.1, this chapter will introduce the detection applications of THz-TDS in the paper and banknote industry,

T. Chen et al., *Terahertz Time Domain Spectroscopy*,
https://doi.org/10.1007/978-981-95-2605-5_8

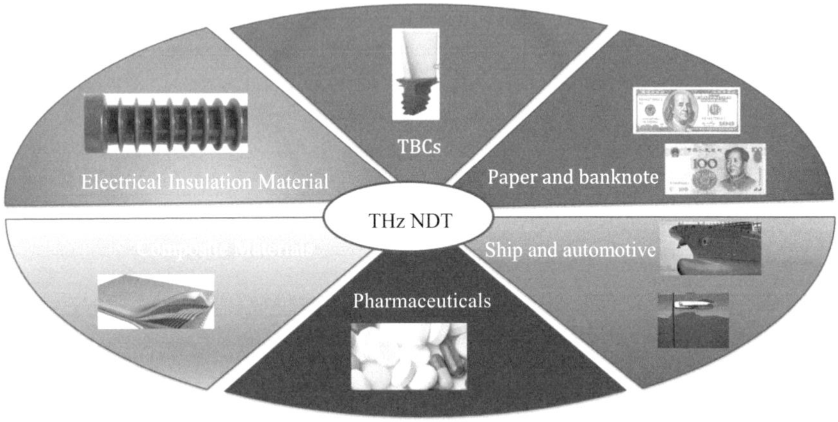

Fig. 8.1 THz-TDS applications in non-destructive testing industries

internal combustion engine thermal barrier coatings, automotive and ship coatings, composite materials, tablets and drugs, electrical insulation materials, etc.

8.1 Paper and Banknote Non-Destructive Testing

Paper and banknotes play an important role in daily life. Banknotes are special currency certificates issued and compulsorily used by the state. Testing the properties of banknotes is necessary and beneficial. Banknotes are a special type of paper made mainly from cotton paper, and banknote testing is an important link in its circulation process. As a new type of testing technology, terahertz spectroscopy has shown great potential in banknote recognition. A joint research team from Tianjin University and the Shanghai Institute of Applied Physics of the Chinese Academy of Sciences used THz-TDS technology to characterize the optical properties of several types of paper and banknotes, with a spectral detection range covering frequencies from 0.5 to 10 THz [1]. Paper and banknotes have significant fingerprint spectra in the terahertz band, which is highly related to the vibration mode of the main component of paper and banknotes—cellulose. Terahertz transmission imaging measurements were also measured. It shows the imaging capability for key anti-counterfeiting labels on banknotes such as watermarks, security printing, and optically variable inks.

Two types of THz-TDS systems were used in the experiment. The traditional THz-TDS system (Advantest, TAS7400TS) obtained low-frequency spectra in the frequency range of 0.5–3.5 THz. The self-made air plasma THz-TDS system based on air plasma generation and detection technology obtained higher frequency information above 2.5 THz. The terahertz absorption spectra of five different banknotes: 5 euros, 5 dollars, 100 ringgits, 100 baht, and 100 yuan are measured. Each absorption spectrum consists of two curves, measured by two THz-TDS systems in the overlapping frequency region. The thickness of the five banknotes used in the measurement is very similar, about 109.6 ± 3.3 um. From the measurement results, the five banknotes have similar terahertz spectral responses, with four significant absorption peaks near the frequencies of 3.05 THz, 5.13 THz, 6.32 THz, and 7.03 THz. In the 2.5~3.5 In the THz frequency overlap region, there is a small deviation between the measurement results of the traditional THz-TDS system and the broadband THz-TDS system of air plasma, which is due to the different terahertz focal sizes of the two THz-TDS systems. As a reference comparison, the absorption spectra of six types of common paper, including sketch paper, red colored paper, filter paper, kraft paper, rice paper, and napkin paper, were measured. All types of paper have similar spectral resonance peaks near the frequencies of 3.05 THz, 5.09 THz, and 6.92 THz, but the relative amplitude of each peak is different. The resonance characteristics of the measured spectra of banknotes and all types of paper are obtained from the vibrational response of their molecules.

The terahertz absorption spectra of banknotes and paper are mainly caused by the response of their cellulose. The absorption spectrum of cellulose was measured. Combined with the results of density functional theory calculations, the cause of the resonance peaks in the absorption spectrum can be explained. The measured peak at 3.08 THz is related to the calculated peak at 3.06 THz, which is caused by the out-of-plane bending vibration of the pyranose ring of the central chain. The measured peak at 5.11 THz corresponds to the vibrational mode calculated at 5.45 THz, which is mainly caused by the vibration of the hydroxymethyl group of the origin chain and the translational vibration of the hydroxymethyl group of the central chain. The modes calculated at 7.02 THz and 7.14 THz produce the measured peak at 6.99 THz, both modes come from the same vibrational pattern. Theoretical calculations show that the terahertz response of cellulose mainly comes from the overall vibration of the chain, and the central chain and source chain usually show different vibration forms under specific terahertz modes.

The THz-TDS transmission imaging of a 100 Yuan RMB banknote was obtained by point-by-point pixel scanning Because higher frequencies correspond to shorter wavelengths and smaller focal points, image quality and resolution can be improved with increasing frequency. At 1.0 THz, the watermark image is almost invisible, and details can be clearly distinguished at 1.5 THz or higher frequencies. As the frequency increases, the image quality improves, and more details can be displayed at higher frequencies. Due to the presence of metal materials and ink materials in the anti-counterfeiting area, the transmittance of the two security lines and the golden "100" letters in the optical image is lower than other places on the banknote. Near the right security line, horizontal stripes invisible in the optical image are clearly visible in the terahertz image. The portrait on the left corresponds to the watermark in the white area, which is almost invisible in the optical image; the portrait on the right corresponds to the red portrait printed on the banknote. The watermark image shows different absorption responses at different frequencies. The rest of the banknote shows a relatively blurry image, where some graphic outlines can be seen, but due to resolution limitations and low contrast, some details are missing.

Using THz-TDS imaging technology, invisible metal material security lines and portraits in optical images can be seen, providing distinguishable imaging effects for different banknotes, and some security marks can be found, which is useful for non-destructive identification of the authenticity of banknotes.

8.2 Thermal Barrier Coating Thickness and Defect Detection

The aerospace manufacturing industry can be regarded as the crown of modern industry, and the aerospace engine is considered the jewel in the crown. With the continuous increase in the temperature and pressure of the combustion chamber of the aerospace engine, it is necessary to enhance the high-temperature resistance of the hot-end components of the aerospace engine. As one of the core components of the aerospace engine, high-pressure turbine blades need to withstand high temperatures and harsh loads. High-temperature alloys are widely used to construct engine blades. With the rapid development of aerospace technology, single turbine blade high-temperature alloy material cannot withstand the high-temperature environment of engine operation. Thermal Barrier Coating (TBCs) are applied to the turbine blades of aircraft engines. It can effectively reduce the working temperature of the turbine blade alloy and significantly improve the efficiency and thrust of the engine and thus extend the working time of the engine [2]. The thickness and uniformity of the thermal barrier coating ceramic insulation layer can affect the insulation performance and service life of the thermal barrier coating. In high-temperature environments, as the working time increases, the thickness of the ceramic layer decreases, which will affect the insulation performance of the thermal barrier coating. Therefore, the detection and evaluation of the thickness of the thermal barrier coating is a hot issue that needs to be studied at present. Generally, the methods for detecting the thickness of the thermal barrier coating are divided into destructive detection and non-destructive detection. Destructive detection methods mainly include metallographic methods and scanning electron microscope methods, while non-destructive detection methods mainly include acoustic emission detection, ultrasonic detection technology, infrared detection, eddy current detection, microwave detection, radiation detection, etc. Terahertz waves have high penetration and resolution better than traditional non-destructive detection methods for thermal barrier coatings. In recent years, THz-TDS technology has made great strides in non-destructive detection of thermal barrier coatings.

8.2.1 Vertical Incidence Detection of Single Layer Thermal Barrier Coating Thickness

T. Fukuchi first proposed using the reflection waveform of terahertz waves to determine the refractive index and thickness of the surface coating of the thermal barrier coating [3]. The refractive index of the ceramic layer was extracted based on the reflection resonance peak, and the thickness of the sample was calculated based on the time delay between the two pulse echoes from the front and back surfaces of the ceramic layer. Considering the impact of surface roughness in practical applications, the reflection coefficient is compensated to obtain more accurate thickness detection results [4]. This method can only obtain the refractive index of the top layer material, and the increase of oxides will cause multiple reflection resonance peaks in the thermal barrier coating to overlap, which is not conducive to the separation of reflection signals from each material layer and has a negative impact on the extraction of the refractive index.

In 2021, a research group from Capital Normal University used transmission THz-TDS technology to measure the thickness of the thermal barrier coating of gas turbine blades [5]. The optical properties of yttria-stabilized zirconia ceramics (Yttria-stabilized zirconia, YSZ) were obtained using transmission THz-TDS system, and the thickness of the thermal barrier coating was measured through reflection experimental setup. In the reflection thickness measurement, considering the impact of dispersion-induced time-domain pulse broadening on the coating thickness, the peak-to-peak thickness and valley-to-valley thickness of the coating reflection signal were calculated separately. The thickness measurement results meet the requirements for process quality evaluation, further elaborating the application of terahertz nondestructive detection technology in insulation layer thickness measurement.

Firstly, the incident terahertz wave vertically penetrates the sample, obtaining the terahertz waveform of the air as a reference and the YSZ ceramic sample. The refractive index of the thermal barrier coating increases linearly with the increase of frequency, and the absorption coefficient increases with the increase of frequency. The thermal barrier coating does not have obvious absorption characteristics. After calculation, the average refractive index of the thermal barrier coating in the effective frequency range is 4.10, and the average refractive index calculated from the time-domain spectroscopy signal is 4.07. The refractive index errors obtained by the two calculation methods are both 0.03, indicating the consistency of the refractive index calculation results.

When using reflection THz-TDS system to measure the coating thickness, thermal barrier coating sample consisting only of a metal substrate and a YSZ ceramic coating is selected. When the terahertz wave is incident on the sample, the surface echo can be detected. Part of the incident terahertz wave is reflected from the upper surface of the coating. Part is reflected again from the upper surface of the interface between the metal substrate and the coating, and part is reflected multiple times between the coating surface and the interface. The time interval Δt between adjacent reflections is obtained from the time-domain reflection signal, and the refractive index n of the coating material YSZ is obtained through transmission parameter extraction, which can be used to calculate the thickness of the thermal barrier coating.

In the reflection experiment, the selected thermal barrier coating sample is composed of YSZ ceramic coating and metal substrate. The thickness of the three samples is different, with EB-1 being the thickest and EB-3 the thinnest. The spot diameter on the sample is less than 3 mm, each sample is tested at 6 different positions, the relative position of each sample is the same as the 6 black dots marked on the sample EB-1, and each position is measured 3 times to ensure the stability of the signal.

The average time-domain reflection spectra of different samples are measured. Each time-domain signal includes the top surface reflection signal, the bottom surface reflection signal R1, and the interface reflection signal R2. Due to the secondary reflection of terahertz waves in the sample, R2 has a significant attenuation, and the signal characteristics are not obvious. As the three samples become thinner, the interface reflection signals begin to overlap, and the waveform of EB-3 has obvious aliasing. Different wavelengths of terahertz waves have different dispersion at different depths of the coating. Due to different material preparation conditions, the propagation paths of photon scattering in the coating vary. Therefore, the effective optical thickness of the thermal barrier coating is longer than the geometric thickness of the coating, which is due to the broadening of the terahertz pulse in the time domain.

By extracting the peak-to-peak time interval (p-p value) and valley-to-valley time interval (v-v value) of the top surface reflection terahertz pulse and the bottom surface reflection terahertz signal pulse, the coating thickness of each sample was calculated using the time delay method, and the impact of pulse broadening on the measurement results of the coating thickness was analyzed. Referring to the metallographic thickness measurement results, the maximum thickness difference is 6.6%. This research lays the foundation for achieving quality evaluation of blade coatings, further improvement and optimization of the preparation process of turbine blade thermal barrier coatings. It has practical significance for future online monitoring of turbine blades using terahertz technology.

8.2.2 Non-Perpendicular Incidence Detection of Thermal Barrier Coating Thickness

The traditional method is the reflection THz-TDS measurement under the vertical incidence mode, whose optical system is usually bulky and complex. More importantly, this system requires a semi-transparent semi-reflective beam splitter, which reduces the power of the terahertz wave and affects the sensitivity of the detection. In 2021, the University of Applied Sciences in Germany, in collaboration with Siemens, used non-perpendicular incidence terahertz waves for coating thickness measurement [6]. By extending the analytical method to the general case of non-perpendicular incidence of Terahertz waves, the surface roughness of the coating can also play a role in the determination of the coating thickness. YSZ on a nickel–chromium alloy was used. The measurement data of the thermal barrier coating samples validated the proposed new method and were numerically compared with the COMSOL simulation results. In the study, a THz-TDS system operating in reflection mode was used for measurement, incident at an angle of approximately 30° relative to the sample, and the thickness of the sample's scanning electron microscope (SEM) images was analyzed and compared. In the case of non-perpendicular incidence, the Fresnel equations differ for the polarization state of the incident beam. Therefore, separate calculations must be made for parallel and perpendicular polarization, and then averaged according to the polarization state of the incident beam. For partially polarized states, i.e., linearly polarized incident pulses with parallel and perpendicular polarization components, the refractive index can be calculated by superposition of the two. Figure 8.2 shows a reflection-type THz-TDS setup with angled incidence. The sample is fixed on the scanning platform, and the black coating on the lower half is the graphite layer required for thermal imaging measurement.

Four thermal barrier coating samples of different thicknesses were used for measurement, labeled as "6 mils", "7 mils", "9 mils" and "11 mils". First, all thermal barrier coating samples were imaged with a scanning electron microscope, with the SEM image of the sample labeled "7 mils" shown in Fig. 8.3. From top to bottom in the figure are the resin voids used for metallography, the thermal barrier coating, the metal bonding layer, and the steel substrate. From the SEM images, data such as the thickness, porosity, and roughness of the thermal barrier coating samples can be obtained.

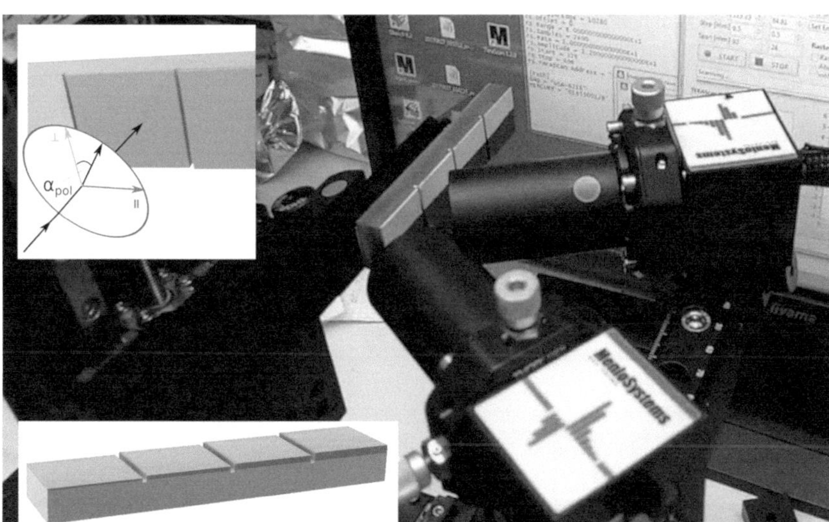

Fig. 8.2 Non-perpendicular incidence reflectional terahertz time-domain spectroscopy system

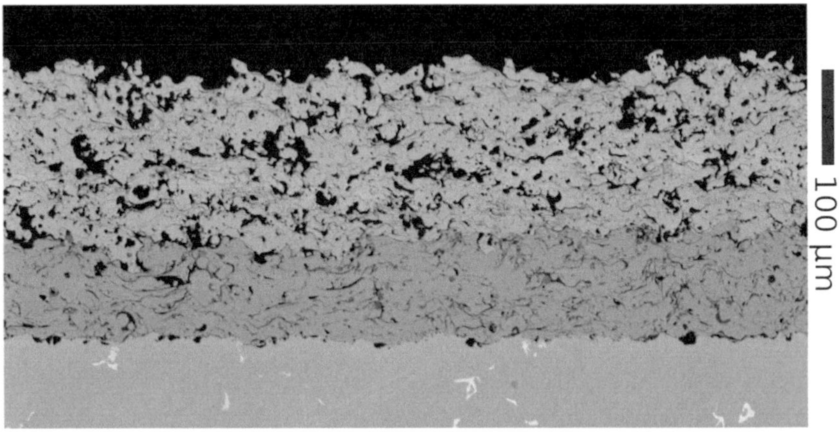

Fig. 8.3 Scanning electron microscopy imaging of thermal barrier coating sample

Combining the time interval of the reflectional THz-TDS signal, the refractive index of the thermal barrier coating samples is approximately 3.872. The roughness correction algorithm can be introduced into the thermal barrier coating thickness detection in combination with theoretical simulation. Table 8.1 shows the thickness calculation results for vertical incidence, non-vertical incidence, and non-vertical incidence with roughness correction. The method of non-vertical incidence with roughness correction has the smallest deviation and the highest accuracy. The absolute error range is between 4 μm and 17 μm, and the relative error range is 1.2%-4.5%.

Table 8.1 Comparison of thermal barriers coating thickness detection errors for different terahertz time-domain signal incidence methods

Sample	Reference thickness (um)	Vertical incidence deviation (um)	Non-vertical incidence deviation (um)	Non-vertical incidence deviation with roughness correction (um)
6 mils	206	+22	+36	+9
7 mils	236	+10	+26	−7
9 mils	330	+19	+42	−4
11 mils	380	+37	+64	+17

8.2.3 Thermal Barrier Coating Surface Microstructure Characterization

The preparation of thermal barrier coatings is a complex thermal, mechanical, and chemical process. The comprehensive insulation performance and lifespan of thermal barrier coatings are closely related to their microstructure, with specific parameters including thermal conductivity, thermal expansion coefficient, elastic modulus, overall stiffness, and fracture toughness. The microstructure is influenced by the properties of the materials, deposition process parameters, and operating environment. Air and combustion gases can more easily penetrate thermal barrier coatings with high porosity, but due to the porous structure, stress mismatches caused by thermal expansion during temperature changes and cycles can be compensated. Moreover, as the usage time of the thermal barrier coating increases, the optimized microstructure introduced during the processing will change, enhancing the thermal conductivity of the thermal barrier coating and weakening its strain resistance. Effectively characterizing its microstructure is crucial for monitoring the lifespan of thermal barrier coatings, optimizing performance, and developing new thermal barrier coatings. Terahertz non-destructive testing technology can effectively monitor the porosity, cracks, and stress changes of thermal barrier coating materials [7].

In 2020, Ye et al. proposed a new method to characterize the microscopic features of thermal barrier coatings based on a machine learning algorithm for terahertz spectroscopy [8]. Their research adjusted the microstructure of YSZ atmospheric plasma sprayed thermal barrier coatings by selecting different types of spray powders, distances, and power. Using a vertically incident terahertz time-domain transmission system, they estimated the terahertz characteristics of porous YSZ ceramic coatings, including refractive index, extinction coefficient, and relative time-domain broadening ratio. The trend of terahertz performance changes in YSZ ceramic coatings with different porosity, pore ratio, pore size, and other microstructural features was observed. Using Principal Component Analysis (PCA), they reduced the number of refractive index and extinction coefficient spectral data in the 0.6–1.4 THz range, ensuring that different characteristics could be input with similar weights during the modeling process. They established three types of models for regression analysis, including Multiple Linear Regression (MLR), Back Propagation (BP) neural network, and Support Vector Machine (SVM). Based on the contribution rate of feature vectors, they selected the first principal component of refractive index spectral data and the first two principal components of extinction coefficient spectral data as model inputs. A comparison of correlation coefficients showed that the characterization accuracy of PCA-SVM reached over 95%, outperforming other models.

The THz-TDS system is used to extract terahertz characteristics that represent the microstructural features of thermal barrier coatings. The terahertz characteristics of thermal barrier coatings with processing parameters are extracted to characterize their microstructural features. The accuracy of non-destructive testing measurements depends on whether the attribute parameters are appropriately selected, and different terahertz attributes have different sensitivity levels to different microstructural features. The terahertz signal undergoes multiple reflections on the surface and inside of the thermal barrier coating. From the transmitted reference and sample signals, the frequency domain function can be obtained through Fast Fourier Transform, and further calculations can yield the refractive index and extinction coefficient of the YSZ sample. As the porosity increases, more air, a medium with a lower optical density, is mixed into the YSZ medium with a higher optical density, thereby reducing the refractive index. Various effective medium theories are used to study the relationship between pore shape, refractive index, and porosity; as the pore size increases, the absorption and scattering of the porous scatters also increases, and the extinction coefficient also increases. Considering the increase in scattering with the decrease in wavelength and the configuration parameters of the THz-TDS system, the frequency of 0.6–1.4 THz is selected to extract the refractive index and extinction coefficient. Scattering and dispersion effects will cause broadening of the terahertz pulse waveform, and the microstructural features determine the amplitude of the broadening change. To accumulate more broadening effects to characterize the microstructural features of the thermal barrier coating, the relative time-domain broadening ratio $r = \Delta t_s/(\Delta t_r d)$ of the third transmission is given priority. In fact, all microstructural features theoretically affect the terahertz characteristics of the thermal barrier coating. The refractive index n, extinction coefficient α, and relative time-domain broadening ratio r are selected as inputs for the machine learning model to characterize microstructural features. 80% of the terahertz attribute database is used as the training set to construct BP and SVM models, with the remaining 20% used for validation and prediction. In their research, they used Principal Component Analysis to reduce the dimensionality of terahertz spectral data. They established models using MLR, BP, and SVM algorithms, and evaluated the predictive performance of these models using the correlation coefficient R.

Four typical samples with different thicknesses were selected for study to demonstrate the trend of changes between their microstructural characteristics and terahertz properties. Due to the increase in the number of scattering bodies caused by the optical thin film medium air, the refractive index and extinction coefficient decrease and increase respectively. Among the four samples, the ceramic coating with the highest porosity and largest pore size (40–96 um + 90 mm + 36 kW) has the smallest refractive index value and the largest extinction coefficient value. Due to the scattering effect, terahertz photons follow a longer path during transmission, and multiple scattering also accumulates increased dispersion, and the relative broadening ratio also increases accordingly. Therefore, all these terahertz characteristics can be effectively used to characterize the microstructural features of YSZ ceramic coatings, thereby obtaining model inputs.

To ensure the dimensional similarity of different terahertz characteristics, reduce redundant data, and ensure that the actual contribution rate of each terahertz characteristic is similar, principal component analysis is used to reduce the dimensionality of refractive index and extinction coefficient spectral data. The eigenvalues of the first two feature vectors extracted from the refractive index spectral data are 420.52 and 0.23 respectively; the eigenvalues of the first two feature vectors extracted from the extinction coefficient spectral data are 394.50 and 17.52 respectively. The contribution rates of the first two feature vectors extracted from the refractive index spectral data are 99.89% and 0.06% respectively; combining eigenvalues and contribution rates, the first two principal component factors of refractive index and extinction coefficient are selected as the input for the MLR and SVM models describing microstructural features. For comparison, and to verify the accuracy of these methods, the correlation coefficients R of the PCA-MLP, BP, and PCA-SVM models are 97.58%, 96.53%, and 99.53% respectively, and the prediction accuracy of porosity and pore size for all models has reached over 95%. The combination of terahertz non-destructive technology and machine learning technology for microstructural feature characterization is effective and feasible. It has a profound impact on the structural integrity assessment of thermal barrier coatings in gas turbine blades.

8.3 Coating Thickness Measurement for Ships and Automotives

The shipbuilding and automotive industries are the engines of the national economy, representing a "1:10" industry, i.e., a 1% increase in automotive industry output can drive a 10% overall increase in the output of all sectors of the national economy. The inspection of the thickness of coatings on ships and automobiles is an important part of quality control for coatings. If the coating thickness is not up to standard, it can lead to insufficient adhesion and reduced protective function. THz-TDS technology is widely used for hull and automotive paint thickness measurement.

8.3.1 Ship Hull Coating Inspection

Organic protective coatings are typically applied to the surfaces of marine or oceanic structures. These organic protective coating systems play a crucial role in the corrosion protection of offshore metal structures. Health monitoring of protective coatings, especially non-destructive testing of local hidden defects, and the detection of possible coating defects and evaluation of coating performance have practical significance for corrosion degradation monitoring. Reliable defect identification can provide timely and effective maintenance, avoiding serious consequences. Several traditional methods are used to characterize paint layers, such as electrochemical techniques, ultrasonic testing, eddy current testing, and infrared thermal wave thickness testing methods. THz-TDS technology can perform non-destructive testing on organic coatings and has great potential in industrial applications.

In 2021, Tu et al. used terahertz pulse imaging technology to study various multilayer organic coating systems, exploring the detection capabilities of protective coating non-destructive testing with terahertz [9]. Measurements were taken on several coating samples with defects, providing a comparison of the measurement results obtained for samples with and without defects. The changes in signals caused by the presence of defects were explained, and structural analysis, quantitative evaluation, and defect identification were carried out. The measurement results show that corrosion defects, paint swelling defects, and paint peeling defects can all be distinguished and identified in conjunction with the causes of the defects.

A THz-TDS system in reflection mode was used to measure different samples. Figure 8.4 shows the THz-TDS signals and scanning electron microscopy imaging results for various coating samples. On each sample, several points were selected from the terahertz test results, and the thickness was calculated using the obtained refractive index. Due to the strong reflection of the metal substrate, the peak reflection intensity of the coating and metal interface is significantly stronger. At the same time, the peak value reflected from the air and coating is also stronger. The peaks produced by the antifouling and connecting ring coatings can also be easily distinguished, despite their small differences. Terahertz waves have a good ability to distinguish various organic coating structures, and the images obtained with signals and scanning electron microscopy imaging results are used for verification.

Another key issue is the reliable inspection of defective coating structures, which directly affects the widespread application of terahertz non-destructive testing technology. During the corrosion and degradation process of protective coatings, their chemical and physical properties change, forming various defects. Several types of local hidden defects exist beneath the coating: paint swelling defects, paint peeling defects, and metal corrosion defects. These defects can reduce the protective performance of the coating system. The main reasons for the occurrence of hidden corrosion defects are insufficient pretreatment of the metal substrate surface before spraying, leading to the presence of residual oxides, rust, etc.; or long-term seawater penetration during service causing chemical corrosion of the metal. Figure 8.5 presents the terahertz time-domain pulse results of a corroded defective coating. The defective sample is sprayed with a layer of anti-corrosion coating before painting, and the rust layer is retained on the metal surface. Figure 8.5a shows the terahertz time-domain pulse signals at eight different points in the X direction. Figure 8.5b, c display the top view and cross-sectional imaging of the defective sample. When testing the sample with terahertz, each pixel of the image corresponds to the peak amplitude of the reflected terahertz pulse in the time domain. By processing the scanned pixels in a specific way, the thickness of the paint layer can be further obtained. If improper substrate pretreatment occurs during the painting process, leading to minor corrosion, it can be quickly detected after the first painting, and remedial measures can be taken in time to clean up the residual oxides on the substrate. In addition, terahertz pulses can obtain uniform distribution information of the drawing from the space of the two-dimensional terahertz image.

The appearance of metal oxides on the substrate will uniformly reduce the paint layer, making the detection results more complex. Even so, corrosion defects can be distinguished from terahertz pulse. Combined with the corresponding pulse function, the corrosion defects under the coating can be quantitatively evaluated. In addition to corrosion defects, research is also conducted on paint swelling defects and paint

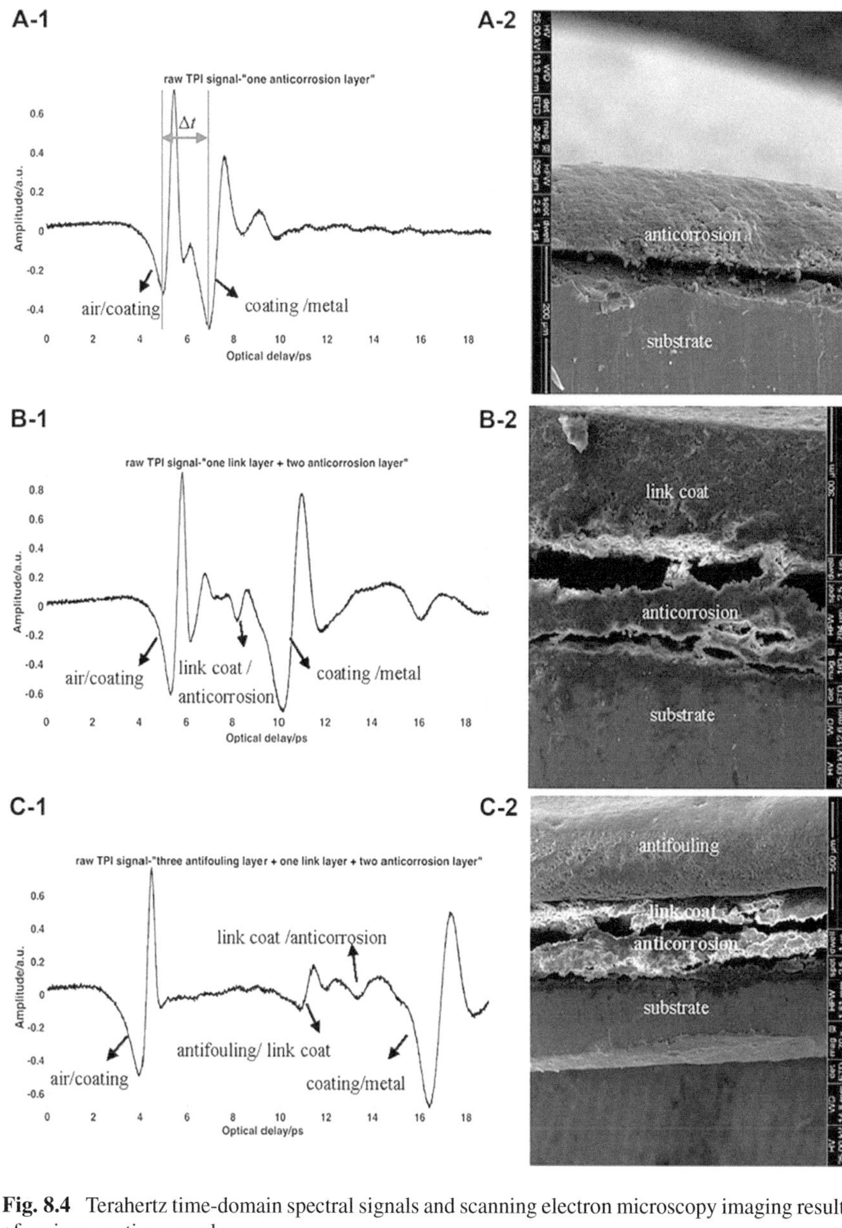

Fig. 8.4 Terahertz time-domain spectral signals and scanning electron microscopy imaging results of various coating samples

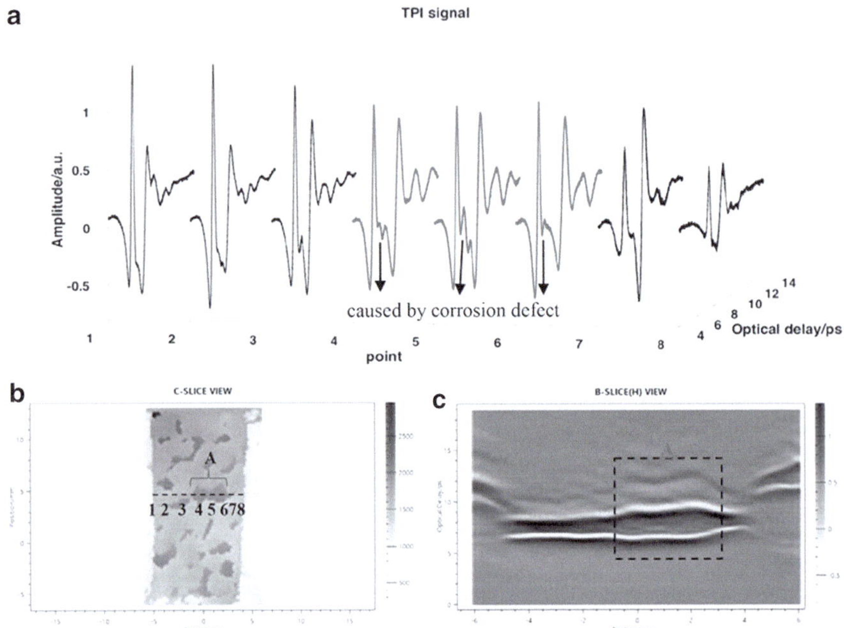

Fig. 8.5 Terahertz time-domain spectroscopy results of corrosion defect samples with anti-corrosion coating layer + metal substrate **a** Terahertz time-domain pulse signals at several points of the sample; **b** Top view imaging of the sample; **c** Cross-sectional imaging of the sample

peeling defects. Traditional techniques are difficult to directly determine whether the separation is between adjacent paint layers or between the coating and the metal interface. But the type of defect is related to the evaluation of coating performance and can be achieved by combining terahertz detection with the cause of the defect. Figure 8.6 shows the terahertz time-domain pulse test results of the coating structure with paint bulging defects and paint peeling defects. The test sample is only coated with anti-corrosion paint and anti-fouling paint layers. Figure 8.6a shows the pulse waveform of the terahertz time-domain pulse of the coating system with bulging defects. The separation occurs within the anti-fouling paint layer and can be identified by type. Paint bulging is usually caused by long-term seawater penetration, which leads to separation between adjacent paint layers. On the other hand, Fig. 8.6b is the terahertz time-domain pulse signal of the coating system with separation defects between the coating and the substrate, mainly due to the decrease in adhesion between the coating and the metal substrate. When bulging defects or paint peeling defects occur, they are very similar to the appearance of the coating. Through the results of terahertz pulse signal detection, these two types of defects can be quickly and clearly distinguished and identified, which is helpful for further evaluation of coating performance.

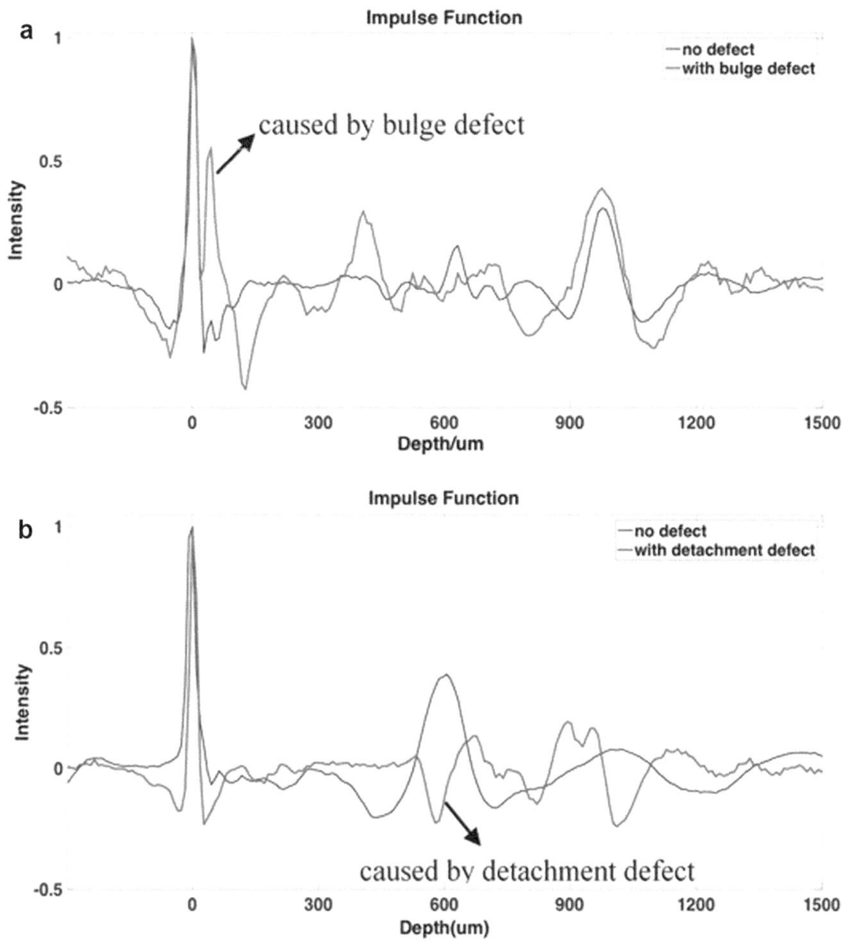

Fig. 8.6 Terahertz wave test pulse waveforms of various coating separation samples **a** Protrusion defect; **b** Detachment defect

8.3.2 Car Paint Thickness Measurement

The automotive industry is highly competitive and requires the strictest online quality control processes. Currently, the body coating thickness measurement used worldwide requires contact measurement, which may damage the car's paint and clear coat; it measures the thickness of the entire coating instead of each layer. It is only effective for metal substrates and ineffective for plastic parts, such as rearview mirrors and bumpers; and some methods are only effective for flat surfaces. The current detection methods have shortcomings, and THz-TDS non-destructive testing methods can be applied to car paint coating thickness testing.

Cambridge University and Aston Martin have developed a new terahertz sensor for real-time measurement and quality assessment of automotive paint thickness based on THz-TDS technology [10]. They use terahertz pulse imaging as a new tool to measure the thickness and quality of up to four layers of automotive paint on metallic and non-metallic substrates. Using a rigorous one-dimensional electromagnetic model for the propagation of terahertz waves in multilayer media, simulated reflection time-domain terahertz waveforms can be calculated. By iteratively guessing the initial thickness and using the least squares method to minimize the sum of the residuals between the measured and simulated waveforms, the thickness of each coating can be extracted, as shown in Fig. 8.7. The B-scan image shows single waveforms as a function of the vertical scan position, with color lines in the B-scan image representing the interfaces between different paint layers. The first, second, and third positive reflection peaks come from the air-clear coat, primer-primer1, and electrophoretic paint-metal boundaries, respectively.

Terahertz time-domain pulses can not only measure average thickness but also resolve the uniformity of coating thickness across the entire sample surface to determine the uniformity of the paint layer and detect paint layer defects. Terahertz measurements are conducted within a two-dimensional area, producing a thickness

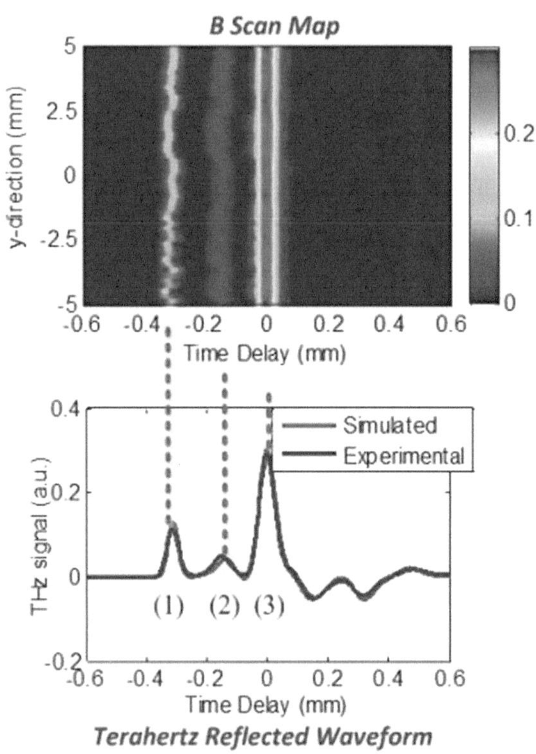

Fig. 8.7 Relationship between car paint b-scan image and terahertz time-domain spectral signal

map as a function of position, which can clearly show significant thickness variations across the entire surface, which is a key factor in determining paint layer quality.

Volkswagen has adopted a solution from das-Nano for non-destructive measurement of automotive coating thickness. das-Nano has developed THz-TDS technology and proprietary algorithms to non-destructively measure the thickness of each coating on the car body. It can provide car manufacturers with previously inaccessible information about the painting process and suggest a 15% reduction in overall process costs by reducing materials, energy, and defective parts. The system allows for quick and non-contact inspection of materials, real-time thickness measurement of each painted and wet-dry and cured clear coat layer, and early detection of defective parts. The system is fully automated and self-calibrating, with no need for a stop-and-go calibration system. It can be applied to any type of substrate, including metals, polymers, and composites, and is suitable for both flat and curved surfaces, with accuracy reaching the micron level. It can provide real-time data on the thickness of each paint layer, control and monitor the painting process, detect quality errors early and correct them, and cover the entire process. das-Nano's terahertz technology was selected in Volkswagen's international tender and installed the world's first paint thickness detection system in Volkswagen's Navarra factory. Since July 24, 2020, the THz-TDS technology for detecting paint thickness has been successfully fully operational. The automotive industry will continue to rely on these new paint process inspection technologies, and terahertz non-destructive testing can significantly reduce environmental impact. Using a single THz-TDS system for production line vehicles can reduce each car's energy consumption by 50 KWh during the painting process, which means that more than 16 GWh of electricity can be saved each year in the Volkswagen Navarra factory, reducing CO_2 emissions by over 7000 tons annually.

The automation of automotive coating inspection is a major trend. In 2021, robotics giant ABB publicly announced its development work on a coating thickness measurement system using THz-TDS technology, installing the terahertz time-domain sensing measurement system on industrial robots [11]. They studied the light-matter interaction of dry polymer coatings and used advanced new signal processing algorithms to determine the state of the dry coating material. This is the first sensor system for inspecting wet coatings, which can accurately predict the final dry thickness without needing calibration for the type of coating. Figures 8.8a, b show the optical path design of the external and internal sensors of the THz-TDS sensor. Inside a 3D printed shell, two fiber-guided terahertz antennas are placed on a collimated light path system. The bodies of cars are always curved to some extent, and in the paint shop, the sensor head is not perfectly aligned with the detection area. Through collimation design and focusing design, the optical design is optimized to adapt to curved body measurements. The terahertz sensing probe is installed on an industrial robot, as shown in Fig. 8.8c. Figure 8.8d shows a surface map of a car part measured with a terahertz sensing probe, measuring the uniformity of the paint, and clearly identifying defects in the paint in the upper right corner. Figure 8.8e shows the measurement distance distribution from the sensor head to the car part after the terahertz probe is aligned, clearly showing the convex shape of the car body. The measurement accuracy of the terahertz sensing probe is compared with the eddy current thickness measurement method, as shown in Fig. 8.8f, For the measurement of dry coating thicknesses ranging from 7 to 33 μm, the root mean square error of the terahertz measurement is typical 1.1 μm, which is three times the accuracy of the most advanced eddy current thickness gauges. In almost all automotive paint shops, mechanical vibrations are ubiquitous. Since the terahertz time-domain sensing system operates in the time-domain mode, out-of-plane vibrations can cause uneven compression and expansion of the terahertz pulse. It depends on the vibration frequency of the terahertz spectrometer relative to the sampling frequency, which will interfere with the measurement results of the dielectric constant and coating thickness. To obtain the vibration sensitivity of the sensor, the coating thickness is determined on a flat single-layer paint sample, then confirmed on a curved single-layer sample, and finally, multi-layer flat paint samples are used as a function of out-of-plane vibration. Ultimately, the impact of out-of-plane vibrations on multi-layer paint samples on a flat metal substrate can be determined, including the clear coat top layer of the transparent coating, the primer of the middle layer, and the primer layer composed of phosphates. Through custom beam design, industrial robustness is ensured. The sensitivity of the sensor system is three times higher than the most advanced eddy current thickness sensors currently used on dry coatings.

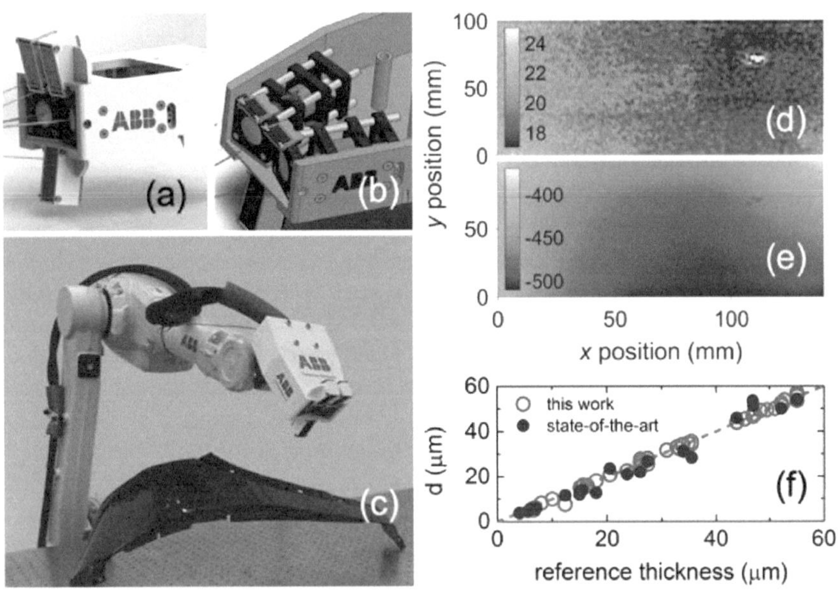

Fig. 8.8 Sensor design and performance **a~c** Terahertz sensor probe design and physical diagram **d** Surface thickness distribution of automotive paint samples coated with primer measured by sensor probe **e** Measurement distance distribution from sensor head to automotive parts **f** Comparison of thickness measurement accuracy between terahertz sensing and Eddy current sensing

8.4 Composite Material Defect Detection

Compared with metal materials, composite materials are lighter and have higher specific strength, better impact resistance and flame retardancy. Composite materials have great application potential in fields such as aerospace, military, energy equipment, and transportation. Commonly used glass fiber composites are composites prepared by winding, manual pasting, and pressing processes with resin as the matrix and glass fiber. During the manufacturing process or within its lifecycle, due to external interference, tiny defects may form internally, and glass fiber composites with defects pose certain safety risks in use. In recent years, many researchers have conducted extensive research on composite materials using THz-TDS systems. Researchers always use reflective THz-TDS system to detect and image layering and other defects in glass fiber composites. They combine wavelet denoising technology to eliminate the interference of water vapor on the measurement results [12]. Reflective tomography is used to analyze the bonding quality of defects, and defect feature time-domain imaging optimization technology is used to improve defect detection capabilities and identification accuracy [13].

In 2020, Zhang et al. proposed a method for using THz-TDS imaging to analyze the thickness of delamination defects in the structural adhesive layer of ceramic-based composites [14]. The terahertz pulse signal is reconstructed using wavelet coefficients, and the thickness of the defect is calculated by the flight time difference between multiple echoes and the base echo. New feature fusion imaging method for detecting delamination areas was proposed. This method first uses THz-TDS for non-destructive testing of the structure of ceramic-based composites. Multiple terahertz detection images with different features are combined to improve image quality and avoid the disadvantages associated with imaging with different features.

In traditional terahertz imaging algorithms, the time-domain waveform of each spatial point of the sample is first obtained, and then the corresponding frequency-domain waveform is obtained using the Fourier transform. Images are reconstructed pixel by pixel based on the maximum and minimum peak amplitudes in the time-domain or frequency-domain waveforms. A method of image fusion based on quantitative analysis of multiple echoes and fusion of various imaging features is proposed. The terahertz signal of the ceramic-based composite structure measured by the THz-TDS system is shown as the black curve in Fig. 8.9. For a ceramic-based composite structure with a delamination defect thickness of 500 μm, the delamination thickness can be obtained directly from the positions of multiple echoes. When the thickness of the delamination layer is 200 μm and 350 μm, the multiple echo signals are usually mixed with the base echo signal, as shown in Fig. 8.9a, b. The continuous wavelet transform method can be used to separate the terahertz multiple echo signal and the base echo signal and obtain the accurate position of the peak. The red curve in the figure represents the waveform after continuous wavelet transform processing, the reflection peak becomes narrower, and the small interface peak is enhanced. For a delamination defect thickness of 500 μm, the separation effect is enhanced using the continuous wavelet transform compared to the result obtained directly based on multiple echoes. When the preset defect thickness is 500 μm and 350 μm, the calculated thicknesses are 513 μm and 378 μm, respectively, which are in good agreement with the actual values. When the thickness of the delamination defect is 200 μm, the detection accuracy is lower because the multiple echo features are annihilated by the base echo. When the defect thickness is greater than 350 μm, the thickness of the defect can be accurately obtained using the time delay between multiple echoes and the base echo, and the quality of the adhesive layer can be accurately assessed.

Images of delamination defects with thicknesses of 200 μm, 350 μm, and 500 μm were obtained using different imaging methods. Three imaging methods were chosen in the study to select specific images. Choosing different feature peaks led to different imaging effects. Taking the imaging results of the sample with a thickness of 500 μm as an example, as shown in Fig. 8.10b, e, minimum imaging can effectively identify the edges of delamination defects, but there is a higher misjudgment rate in normal areas. In Fig. 8.10c, f, although amplitude imaging has better recognition ability, it causes more noise. After continuous wavelet transformation, the overall smoothness of the image has improved, but the edge recognition ability is still poor. Multi-echo amplitude imaging is significantly affected by defect thickness. For delamination defects with a thickness of 500 μm, this method can obtain good imaging results, as

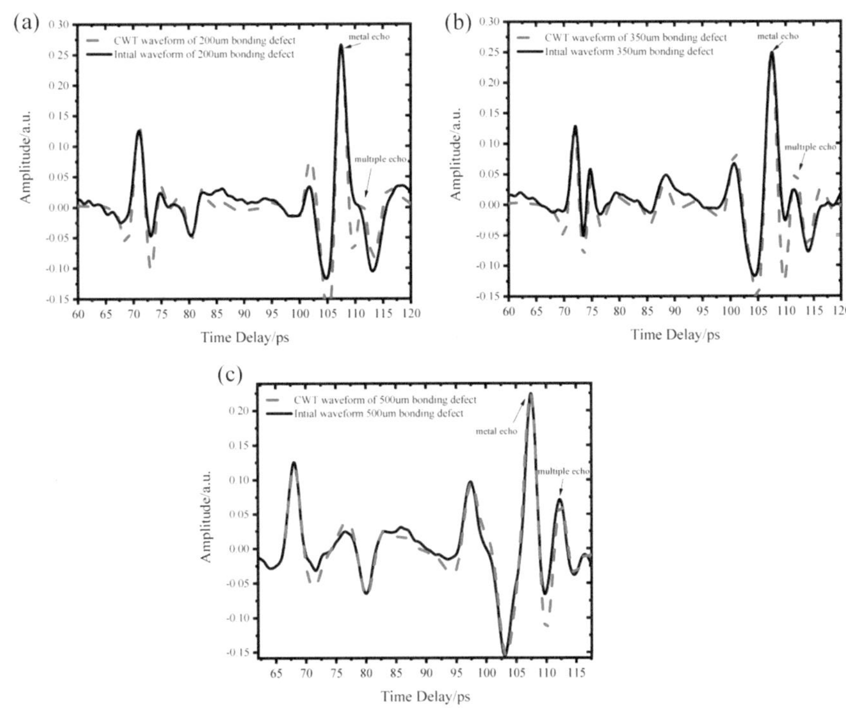

Fig. 8.9 Time-domain waveforms of different defect thicknesses reconstructed by continuous wavelet transform **a** 200 um thickness defect **b** 350 um thickness defect **c** 500 um thickness defect

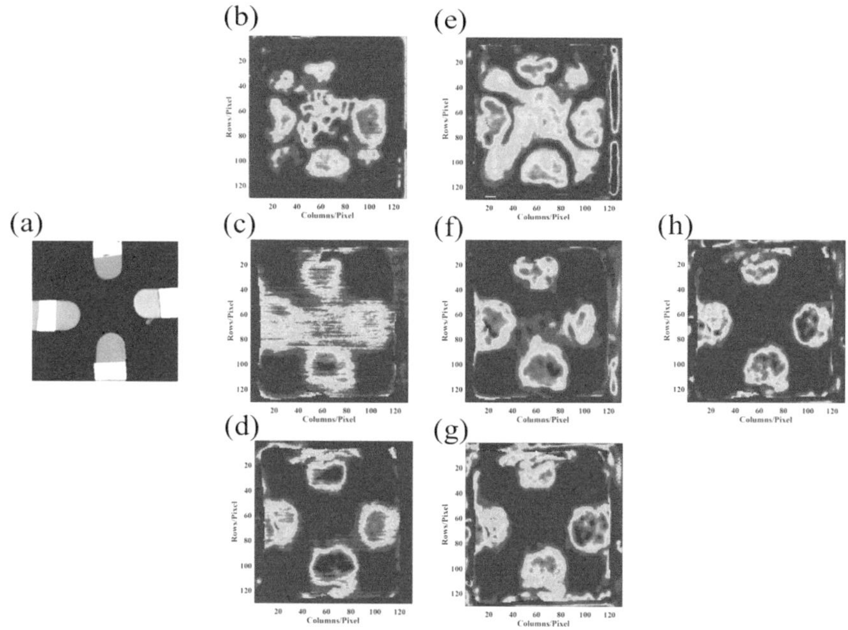

Fig. 8.10 Comparison of different imaging methods for a 500 um thickness delamination defect **a** Optical image of sample defect **b** Minimum imaging method **c** Amplitude imaging **d** Multi-echo amplitude imaging **e** Minimum imaging after continuous wavelet transform **f** Amplitude imaging after continuous wavelet transform **g** Multi-echo amplitude imaging after continuous wavelet transform **h** Feature fusion imaging

shown in Fig. 8.10d, g. Compared with minimum imaging, the misjudgment rate of non-defect areas in feature fusion imaging is greatly reduced, as shown in Fig. 8.10h. Compared with amplitude imaging, it improves the recognition of delamination edges and the contrast of non-defect areas.

Further use of the binary method to calculate the defect area of each slice in the image, thereby verifying that feature fusion imaging has higher accuracy than physical imaging. When the delamination defect thickness is 200 μm, the total error is 139.0 mm^2, the percentage error is 10.9%, when the delamination defect thickness is 350 μm, the percentage error is 11.4%. When the delamination defect thickness is 500 μm, the total error is 120.6 mm^2, the percentage error is 6.78%. Feature fusion imaging has higher stability and accuracy in characterizing delamination defects and calculating delamination defect areas. The delamination defect areas calculated by different imaging methods have large differences. Fusion imaging is superior to other imaging methods for delamination effects of defects of different thicknesses and has good recognition ability for defect areas.

8.5 Non-Destructive Testing in Pharmaceuticals

THz-TDS technology has been widely used for in-depth crystallography and polymorphism studies of drug molecules, as well as identification of drug hydrate forms. The vibrational rotational energy levels of large drug molecules are in the terahertz band, with obvious absorption peaks in the terahertz spectrum, such as levocetirizine with obvious absorption peaks at 1.71 THz and 3.78 THz, and loratadine with obvious absorption peaks at 1.71 THz and 2.91 THz. The terahertz absorption spectra of the two drug molecules are shown in Fig. 8.11 [15].

8.5.1 Drug Coating Thickness Detection

Most raw material drugs are stable in their crystalline forms, but many are not easily soluble in water, which limits their bioavailability. In pharmaceutical production, the thin film coating on solid oral dosage forms, especially tablets, is usually the final step in the secondary manufacturing value chain. Coatings are typically applied to cores, such as tablets or granules, and in some cases can be applied to small particles like crystals. To promote film formation and processing, coating formulations usually contain polymers and additives, such as plasticizers, anti-adhesives, surfactants, and colorants. Coating processes are widely used to ensure uniform color, light resistance, and taste masking of dosage forms. Functional coatings can be used to mask the taste or smell of a product, protect the raw material drug from the acidic environment of the stomach or gastric mucosa from aggressive raw material drugs. It prolongs the release of raw material drugs. Active coatings contain raw material drugs in the coating. They are used to achieve different fixed dose combinations, prevent interactions between different drugs, or combine different release behaviors into a single solid dosage form. The thickness and integrity of the coating are particularly important for functional coatings. A minimum thickness and absence of cracks are needed to ensure the gastric resistance of the dosage form. When the coating has a certain target release curve and rate, the thickness of the coating determines the dissolution rate of the drug. In active coatings, when the distribution of the drug core is uniform and the film density is constant, the amount of raw material drug in the coating is directly related to the coating thickness. Dosage forms with target release profiles are very sensitive to cracks and other defects. Such defects in the coating may lead to dose dumping, spontaneous release of the drug causing possible toxic side effects. Since the coating process is essentially random, the coating thickness of the final product inevitably follows a specific distribution, so the uniformity of the coating is also important.

Most pharmaceutical excipients exhibit no distinct spectral features and are semi-transparent in the terahertz frequency range. Terahertz radiation can thus be utilized for imaging pharmaceutical dosage forms, enabling non-destructive depth-resolved imaging through THz-TDS. When there is an abrupt change in the refractive index at

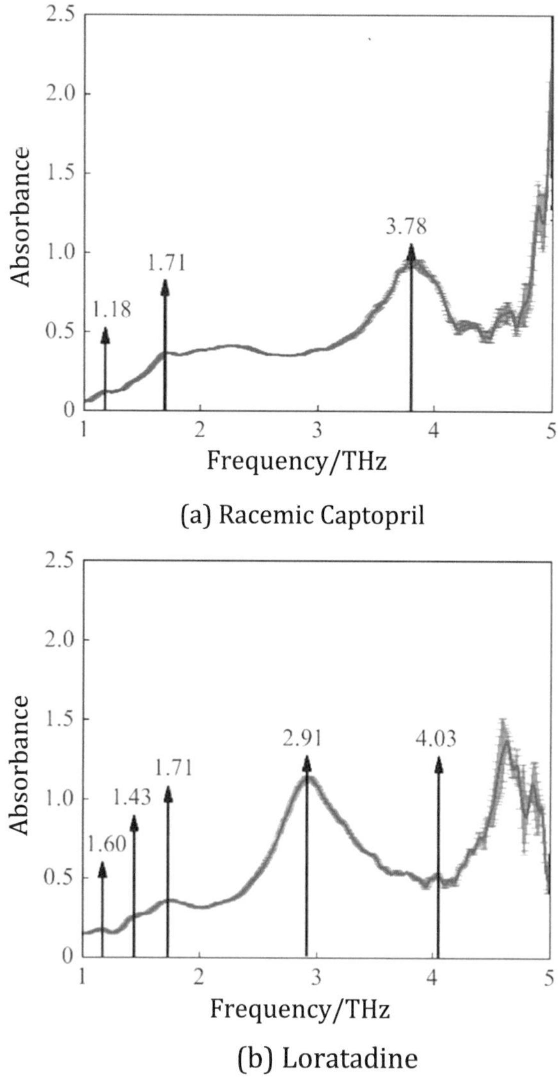

Fig. 8.11 Terahertz absorption spectra of pharmaceutical molecules **a** Racemic captopril **b** Loratadine

each coating interface or between the drug particles and the excipient matrix, a portion of the THz pulse is reflected to the detector, as illustrated in Fig. 8.12a. Figure 8.12b presents a representative time-domain waveform acquired from the measurement. The thickness of a thin film coating is determined by the time delay between adjacent reflection peaks in the processed signal. This method of thickness determination is not adversely affected by imperfect deconvolution settings, but instead requires a refractive index discontinuity between adjacent layers.

By scanning the tablet surface along its geometrical profile, a three-dimensional coating thickness map can be reconstructed, as shown in Fig. 8.12c, where the x-

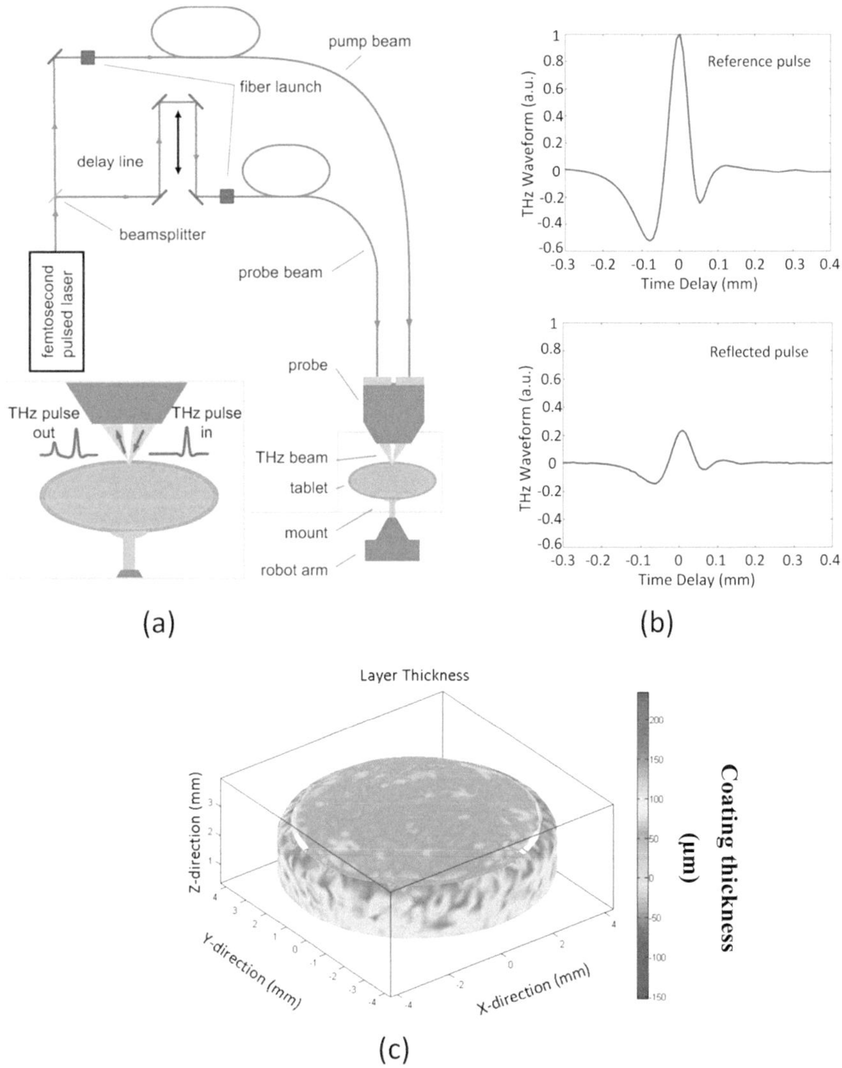

Fig. 8.12 **a** Schematic diagram of terahertz pulse imaging **b** Original terahertz waveform of reference mirror and tablet sample **c** Coating thickness image of biconvex tablet

and y-axes represent lateral dimensions, and the z-axis corresponds to time delay or depth. The lateral resolution is limited by the wavelength of the THz radiation—approximately 150 μm at 2.7 THz and 400 μm at 1 THz—while the depth resolution is constrained by the THz pulse duration. Practically, a lateral resolution of 150–250 μm can be achieved in tablets using a frequency range of 1.5–2.7 THz, with a depth resolution of approximately 30–40 μm.

Numerical fitting of the measured data to a THz electromagnetic propagation model using the known refractive index and extinction coefficient of the materials can significantly enhance depth resolution. In addition to coating thickness, other useful THz parameters include the peak electric field strength, which provides insights into surface roughness and coating density. The THz interface index can indicate physical or chemical transitions at interfaces. However, measurements at sample edges—such as embossed regions, fractures, or highly rough surfaces—may lead to significant scattering and consequently unreliable results. Nonetheless, due to the longer wavelengths of THz radiation compared to infrared or visible light, scattering effects are generally reduced in THz imaging.

The refractive index of excipients used in pharmaceutical surface coatings is critical for accurately determining coating thickness. This refractive index can be estimated from the intensity of the first reflected THz pulse using the Fresnel equations. As a reference, the researchers have published a database of refractive indices in the terahertz range for commonly used pharmaceutical coating excipients [16], as shown in Table 8.2.

8.5.2 Drug Coating Strength Testing

For pharmaceutical companies, when developing tablets, the risk of degradation under a range of conditions must be considered. If appropriate non-destructive tools are not used, these risks will have to be considered in degradation testing. There are many advantages to using terahertz waves for drug analysis. One of them is the analysis of coating thickness using a special measurement principle. Researchers from the Japanese pharmaceutical company Astellas have used terahertz detection technology to study the sources of cracks in the drug coating layer [17]. Under high temperature conditions, in samples where cracks appear in the coating layer, the film surface density (FSD) and interface density difference (IDD) analyzed by terahertz waves are often lower. Therefore, under high temperature conditions, the low density of the coating layer and the unclear boundary between the coating layer and the tablet core are related to the formation of cracks in the coating layer. The index obtained by multiplying FSD and IDD may be used to predict the initiation of cracks in the coating. The film coating strength index ($FCSI = |FSD \times IDD|$) is defined to verify its relationship with crack formation. The smaller the FCSI value, the higher the risk of causing cracks. Figure 8.13 shows the film-coated tablets during and after the accelerated degradation test for 80 min. The coated tablet with the lowest FCSI showed cracks 60 min after the start of the test, while no cracks were found in the tablet with the highest FCSI. As shown in Fig. 8.13, cracks in the coating layer can be clearly identified in the photos taken during the test period from 0 to 80 min. This result indicates that the appearance of cracks can be correctly predicted based on the FCSI value of the film-coated tablets, including tablets with and without cracks in the accelerated degradation test.

Table 8.2 Spectral refractive indices of common excipients for pharmaceutical coatings

Excipient material	Refractive index
Acacia spray dried	1.71 ± 0.02
Acesulfame	2.05 ± 0.01
Avicel PH101	1.70 ± 0.03
Avicel PH102	1.77 ± 0.03
Avicel PH200	1.67 ± 0.03
Avicel PH302	1.76 ± 0.02
Avicel RC581	1.76 ± 0.01
Calcium carbonate	2.13 ± 0.03
Calcium phosphate	2.50 ± 0.01
Calcium phosphate dibasic anhydrous	3.25 ± 0.05
Carboxymethyl cellulose	1.76 ± 0.03
Carboxymethyl cellulose sodium	1.73 ± 0.01
Carnuba wax	1.77 ± 0.01
Confectioners' sugar	1.71 ± 0.005
Cornstarch	1.62 ± 0.01
Dextrose	1.74 ± 0.01
Hydroxypropyl cellulose	1.47 ± 0.01
Lactose anhydrous	1.69 ± 0.005
Magnesium hydroxide	1.78 ± 0.01
Magnesium oxide light	1.44 ± 0.005
Methyl paraben	1.66 ± 0.02
Magnesium stearate 1	1.37 ± 0.005
Magnesium stearate 2	1.34 ± 0.005
Magnesium stearate anhydrous	1.34 ± 0.005
Polyvinylpirrolidone K30	1.60 ± 0.01
Povidone	1.56 ± 0.005
Pregelatinised starch	1.65 ± 0.01
Silicon dioxide colloidal	1.26 ± 0.01
Sodium bicarbonate	1.97 ± 0.02
Sodium carbonate	2.01 ± 0.01
Sodium carboxymethyl	1.78 ± 0.01
Sodium lauryl sulphate	1.68 ± 0.01
Sodium starch glycolate	1.81 ± 0.01
Stearic acid	1.53 ± 0.02
Sucrose	1.83 ± 0.01
Sugar tab	1.70 ± 0.01
Tartaric acid	1.85 ± 0.03
Titanium dioxide	2.29 ± 0.02

(continued)

Table 8.2 (continued)

Excipient material	Refractive index
Xyloitol	1.76 ± 0.03
Xyloitol 300	1.83 ± 0.03

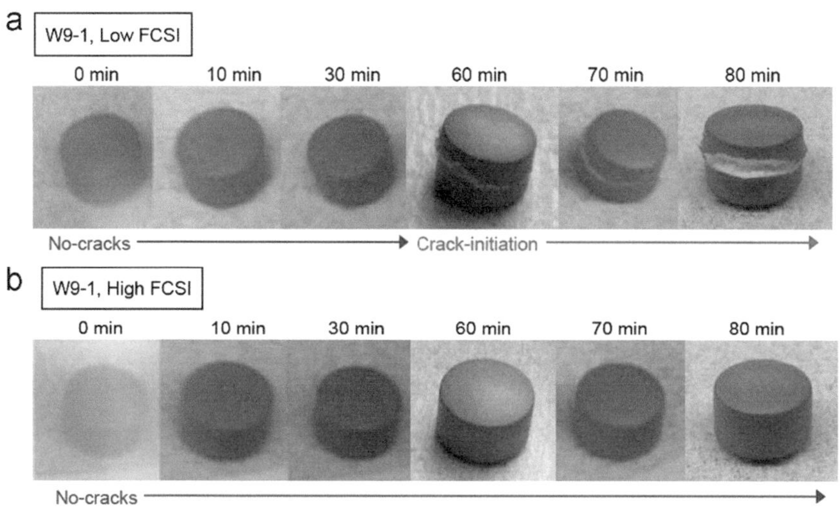

Fig. 8.13 Appearance of film-coated tablet samples before and after degradation experiment

Figure 8.14 shows the relationship between FCSI and crack formation. The results show that there is a good correlation between the FCSI value and the occurrence of cracks. Tablets with smaller FCSI values have a significantly higher rate of crack formation than those with larger FCSI values. These results support the research on non-destructive prediction of crack formation using THz-TDS technology.

Fig. 8.14 Relationship between FCSI value and crack formation

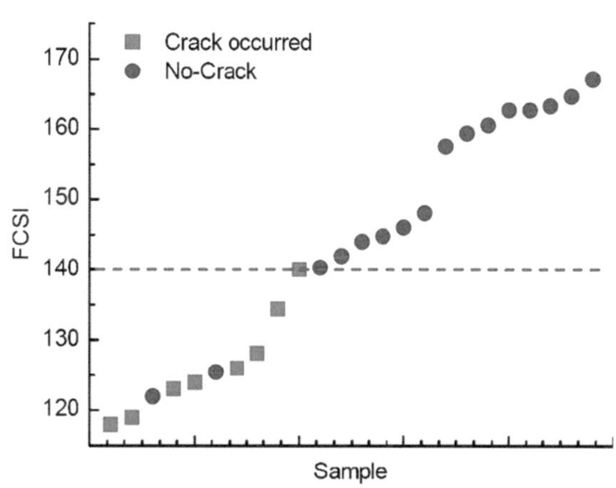

8.6 Electrical Insulation Material Testing

The safety of the power system is related to the safety of the national economy and people's livelihood and has significant strategic significance. In recent years, the safety threats to the power system from both internal and external sources have increased, presenting characteristics such as a high proportion of renewable energy and diversified loads of power equipment access, and the power grid security situation is severe. Building a more perceptive, responsive, defensive, resilient, collaborative, and learning resilient power grid has become the goal of power grid construction. There is urgent need to develop various non-destructive testing of power materials. Non-destructive testing of metal materials is already mature, such as ultrasound and eddy current technologies have been fully applied to metal material testing. Non-metallic materials such as insulators and insulating rods are used to close or open high-voltage isolating switches, install and remove portable grounding wires, and are used for measurement and testing. If there are defects such as delamination, voids, and slag in the insulating rod, it is easy to generate high-voltage discharge, which can then cause electrical fires. Currently, there are few methods for non-destructive testing of non-metallic materials in the power industry. Terahertz waves can penetrate non-metallic insulating materials and can be used for non-destructive testing of electrical insulating materials.

8.6.1 *Interface Air Gap Detection of Composite Insulators*

With the intensification of environmental pollution, composite insulators with anti-pollution flashover characteristics are used in outdoor transmission lines. Internal defects of composite material insulators often lead to failures and result in serious accidents. The common online detection methods include voltage and partial discharge measurement, infrared temperature measurement, ultrasonic measurement, microwave technology, etc. Due to the limitations of processing and preparation technology, there may be air gaps in the interface of composite insulators. After long-term operation of insulators with poor performance, erosion defects will occur at the interface, and the transmission rules of terahertz waves in interfaces with different characteristics are also different. When there are air gaps and erosion defects at the interface, it will affect the reflection and scattering process of terahertz waves, which will cause changes in the amplitude and waveform of the reflected wave. It enables qualitative and quantitative detection of interface defects. Composite insulators are usually composed of core shafts, sheaths, shed skirts, end fittings, and grading rings. Some insulators, such as column insulators and composite cross arms, fill polyurethane in the core shaft, and the sheath and shed skirt generally use silicone rubber, and the core shaft uses epoxy resin. Air gap defects are the most common type of defect, which may form at the interface or inside the sheath and core shaft. Connect the silicone rubber board and the epoxy resin board together, foam and solidify the polyurethane in the mold connected to the epoxy resin board, simulating the multi-layer structure of the insulator. Groove the fiberglass epoxy, simulating air gap defects.

In 2020, Zhang et al. used a reflective THz-TDS non-destructive testing method to evaluate the interface characteristics of composite insulators [18]. Based on the transmission spectra of terahertz waves in silicone rubber plates and epoxy glass fiber plates, the average refractive indices of silicone rubber and epoxy glass fiber plates were calculated to be 1.907 and 2.220, respectively. First, the air gaps present at the silicone rubber-glass fiber interface were detected. Different thickness grooves were prepared on the epoxy glass fiber plate using a slotting process, and then a silicone rubber plate was overlaid to simulate the air gaps present at the interface. During measurement, the distance from the probe to the surface of the silicone rubber was kept constant, so the reflected waves from the first surface of multiple samples coincided. When air gaps were present, the reflected waveform exhibited a double feature of phase inversion and amplitude increase. The thinner the sheath, the larger the amplitude of the reflected wave at the interface, and the greater the difference in the peak values of the reflected waves obtained with and without air gap defects. In actual testing, a single layer of silicone rubber with a thickness of 6 mm was measured, and the reflection spectra curves of the silicone rubber-epoxy glass fiber interface at different air gap sizes were obtained.

The difference between the actual waveform and each curve in the curve set was then calculated, and the defect size corresponding to the smallest difference was the measured size. When the air gap was small, the measurement error of the terahertz wave was large. For example, when the air gap thickness was 0.071 mm, the error was about 15.5%, but when the air gap thickness was greater than 0.264 mm, the measurement error was within 10%, and the overall measurement accuracy was about tens of micrometers. Next, the erosion defects present at the interface were studied. The sample was placed on an optical platform that could adjust the rotation angle minutely, and the terahertz transmitter and receiver were aligned with the insulator cylindrical sample, and the sample was rotated counterclockwise from the defect starting point. During the rotation, the time-domain waveform changes were observed, and when the waveform began to show abnormalities, it was taken as the starting position of the defect, and when the waveform returned to normal, it indicated a normal area. After a 35.6° rotation, 179 data points of terahertz reflection signals were obtained, as shown in Fig. 8.15. The reflected waveforms in the red circle in Fig. 8.15a do not coincide significantly, while they coincide well in other positions. As can be seen from Fig. 8.15b, there are significant differences in the reflected waves at different angles. When erosion is present at the interface, roughness, brittleness, and other phenomena will occur inside the interface, and the conductivity, dielectric constant, and other characteristics of the material at the erosion site will change significantly. When terahertz waves incident on the erosion site, significant absorption loss and scattering occur, and the energy and phase of the reflected waves detected are significantly different from those at the air gap. Both the silicone rubber and the core rod at the interface have erosion, and the reflection and transmission characteristics of terahertz waves on the surfaces of both materials are different from those of

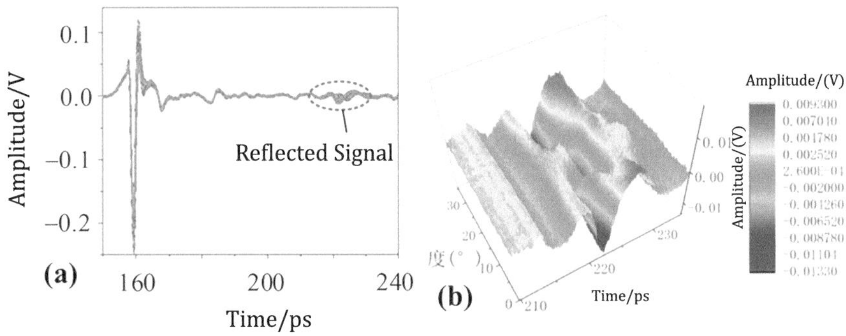

Fig. 8.15 Line scan test reflection waveform data. **a** Interface reflection wave scatter plot; **b** Three-dimensional plot of reflection wave data

undamaged materials. The sum of the squares of the differences between the reflected waves in the defect center area and the normal area waveforms and the sum of the absolute values are greater than in other areas, and there is a certain symmetry with the defect as the center. Reflective terahertz time-domain signal surface scanning imaging was performed on the sample with interface erosion, and the detection area was a 15 mm × 20 mm area around the defect. The sum of the squares of the peak differences at the interface was used as the vertical coordinate to generate a two-dimensional rainbow map. The defect is a penetrating strip, which corresponds well with the actual erosion area. The color depth is greatest in the center of the defect and gradually decreases towards both sides. The results show that THz-TDS technology can evaluate the distribution area and degree of damage of composite insulator erosion.

In 2021, Jiang et al. designed and established a terahertz scanning system that can be used for flat samples and cylindrical samples. They used THz-TDS technology and sample scanning systems to detect defects in composite columnar insulators [19]. There are two air gap defects in the center of the sample, which overlap on the path of the terahertz beam. One of the defects has a depth of 0.40 mm and is star-shaped. It is located at the interface of silicone rubber and epoxy resin. The other defect's depth is 0.83 mm and of square shape. It is located inside the epoxy resin plate, 0.3 mm away from the polyurethane. Since the star-shaped defect is within the range of the square defect, samples with no defects, square defects, and two defect positions were selected for measurement.

The propagation speed of terahertz waves in dielectric materials is slower than in air gap defects. When there are defects at the sampling point, terahertz waves will become much weak after multiple reflections and refractions. The reflection pulse at the interface of the insulator itself can be used to diagnose potential defects. If there are defects in the material above the interface, the amplitude of the reflection pulse will significantly decrease. If the interface has defects, the amplitude of the reflection pulse will be greater than that of the defect-free interface.

Selecting the terahertz time-domain signal imaging corresponding to the B silicone rubber and epoxy resin interface, and the E epoxy resin and polyurethane interface, the maximum absolute value, signal power, and envelope area are extracted in turn for imaging. From the results, all three imaging results can detect two defects at different positions. The maximum absolute value imaging is relatively poor in identifying star-shaped defects. This is due to the shape of the defect is more complex and the edge is not in the scanning direction. Considering the similarity level of the imaging results and the actual defects, the performance of signal power imaging is the best, but there are missing parts in the detect of the edge. This is due to the edges of the imaging results are more blurred than the envelope area imaging. On the other hand, the detection effect of envelope area imaging is weak in terms of angle. Envelope area imaging and signal power imaging are both better at identifying defects than maximum absolute value imaging, and these two parameters are superior in terms of defect morphology recognition and resolution.

8.6.2 Insulating Paper Detection

Insulating paper is the main insulating material of power transformers and is closely related to the operational risk of the equipment. The concentration of field strength at insulation defects is the main cause of partial discharge and failure. The mixture of air gaps and foreign substances such as metal and carbon black in the insulating board are common defects in oil-paper insulation. During the production process of oil-paper insulation, due to uneven washing of the bonding solvent and incomplete volatilization, delamination can easily occur to form cavities. Therefore, the surrounding insulating oil cannot penetrate, and the cavity forms a typical air gap defect. In the production of insulating components, such as the wet forming process of special-shaped parts, tiny metals on the mold can mix into the insulating parts to form foreign substances. Micro-carbonized particles produced during the partial discharge process, or micro-carbonized micro-channels, as well as local moisture accumulation caused by factors such as humidity, may cause local field strength concentration in oil-paper insulation, posing a hidden danger to the safe operation of the equipment. Transformer insulation materials are usually checked for foreign substances mixed with insulating boards or molded parts using X-ray methods before leaving the factory. However, it is difficult to detect the type, position, and shape of possible defects in the internal insulation components of the assembled transformer. Currently, industrial tomography scanners can be used to achieve three-dimensional imaging of insulating board defects. But its operation is complex and costly, and X-ray leakage may occur. Scanning electron microscopes can analyze the surface morphology and material composition of insulating board samples, but the observation area is small and cannot detect internal defects of the insulating board.

In 2021, Li et al. used THz-TDS to study the non-destructive testing method of insulating paper defects [20]. Taking interface delamination, metal impurity mixing, and local carbonization traces as typical insulating board defects, they established artificial defect models and tested and analyzed the time-domain and frequency-domain waveform characteristics of terahertz pulse waves propagating in the defect models. The results show that when the covering thickness of the insulating board is less than 5 mm, the position and size of typical internal defects can be accurately obtained based on the amplitude value and delay characteristics of the terahertz time-domain signal. The experiment verified the feasibility of using terahertz technology for non-contact and non-destructive testing of internal defects in insulating boards.

When terahertz pulse wave propagates through insulating oil paper, it encounters defects such as air gaps and foreign objects, thereby altering propagation amplitude and phase characteristics. The reflection of terahertz pulse waves is used to study the detection of these defects. The sample backing is made of metal, and the total reflection characteristics of the terahertz pulse waves on the backing are used to enhance detection sensitivity. The terahertz waveform measured in the glued sample A-1 only shows a clear reflection wave, which is formed by the reflection of the incident wave on the upper paperboard interface. Two terahertz reflection waves are measured in the delaminated sample A-2. The first signal comes from the reflection of the incident signal on the upper paperboard interface. This signal is highly consistent in time delay and amplitude with the reflection signal measured in the bonding area. The second reflection signal occurs at several tens of picoseconds and consists of two reflection waves of opposite polarity. This signal comes from the reflection signal on the upper and lower interfaces of the air gap at the delamination point. By comparing defect samples of different insulating paperboard thicknesses, the delay and amplitude of the reflection wave signal decrease with the increase in the thickness of the insulating paperboard. This is due to the fixed position of the terahertz measurement probe and the sampling stage during the test. Because the optical path difference between the upper insulating paperboard interface of the sample stage and the probe is reduced, the reflection wave delay is shortened, and the amplitude is reduced. The increase in the thickness of the insulating oil paperboard extends the propagation time of the terahertz wave in the paperboard, thereby increasing the signal attenuation and causing a decrease in amplitude.

Copper powder is used to simulate metal foreign objects in the insulating oil paperboard, and the corresponding terahertz waveforms are analyzed. The first reflection signal of the waveform detecting the adhesive and copper powder positions is the reflection peak of the insulating oil paperboard interface, and other pulse waveforms show different patterns. The residual waveform in the glued insulating oil paperboard only includes a clear reflection wave at about 80 ps, which is formed by the reflection of the terahertz pulse wave on the metal substrate. The copper powder insulating oil paperboard B-2 generates pulse reflection waves at the upper and lower insulating paperboard interfaces and the copper powder interface, and a positive peak reflection wave appears at 40 ps, which is caused by the reflection of the terahertz wave on the copper powder.

Keeping the terahertz measurement probe unchanged, tests are then conducted on the oil-immersed area and carbonization trace area of local carbonization trace defects. Terahertz reflection waves generate pulse reflection waves on the upper, middle and lower layer interfaces, carbonization trace interfaces, and metal plate interfaces of the paperboard. Obvious reflection waves are generated at the interfaces of the upper, middle, and lower insulating paperboards and the metal plate. At 15 ps of the terahertz time-domain waveform, pulse reflection waves are generated at the air gap interface and lower interface of the upper paperboard; at 30 ps of the terahertz time-domain waveform, pulse reflection waves are generated at the air gap interface and lower interface of the middle paperboard. Pulse reflection waves are generated in the locally carbonized area. This is due to the refractive index of the insulating paperboard after carbonization is different from the original insulating paperboard, and the terahertz pulse wave is reflected at the carbonization track.

Using the XY scanning platform to perform point-by-point scanning detection of insulating paperboard samples, with a step accuracy of 1 mm, the terahertz time-domain pulse imaging results of different defects insulating oil paperboards can be obtained. According to the above analysis of the waveforms at the delamination and bonding positions, it can be found that there is a difference in the amplitude of the pulse reflection wave at the air gap interface. The minimum amplitude is used for imaging analysis, and the terahertz time-domain image of the insulating paperboard sample is obtained. The imaging effect is obvious, with the brush positions on the left and right sides. Due to the squeezing and wetting phenomenon during the sample bonding process, there is a delamination defect area in the middle, and its signal strength is relatively large. The results show that the delamination defects of the sample can be well detected through amplitude imaging. The terahertz wave is completely reflected at the copper powder, and its pulse reflection wave is a positive peak. Using the maximum amplitude of the pulse reflection waveform reflected by the copper powder for imaging, the time-domain image of the metal foreign object sample is obtained. The imaging result is consistent with the analysis conclusion of the waveform characteristics of the bonding and copper powder defects, and it displays the metal foreign objects in the insulating paperboard. Based on the difference between the local carbonization track and the normal non-breakdown position waveform, the maximum amplitude of the terahertz pulse peak reflected by the carbonization trace is imaged, and the time-domain image of the local carbonization track sample is shown. The imaging results show that the time-domain image can well reflect the information of local carbonization traces and defects in the insulating paperboard.

8.6.3 Inspection of Moisture-Affected Composite Insulation Structures

Multilayer composite insulation is the most widely used structure in high-voltage equipment such as power cables, insulators, and casings. Due to the different materials of each insulation layer, they exhibit different shrinkage characteristics under the stress of the working environment. This may lead to the formation of gaps between each layer, which could further allow moisture to penetrate, causing the insulation structure to become damp. Cable terminal dampness is the most typical phenomenon in multilayer composite insulation structures, especially in hot and humid environments. Due to imperfect sealing, exposure to air, and long-term medium to high frequency vibrations, moisture may penetrate the insulation layer or even the inner layer of the cable terminal. The intrusion of water can lead to more conductive impurities adhering to the composite insulation structure, forming traces of water with higher conductivity. This change in local conductivity can lead to electric field distortion, ultimately causing breakdown faults.

In 2021, Li et al. proposed a non-invasive method for detecting moisture in multilayer insulation structures based on THz-TDS [21]. A three-layer scattering model was established based on the actual structure of the composite cable terminal and theoretical analysis was conducted, revealing its potential application mechanism. An artificial model of the three-layer structure was established, and a damp insulation test was conducted on this model. Terahertz spectroscopy and imaging tests were performed. Frequency domain signals can effectively identify the moisture inside the multilayer composite insulation. The frequency domain imaging results of the sample can accurately determine the geometric parameters of the damp area. Through further time-domain spectroscopy analysis, the thickness of the insulation material affected by moisture can be quantitatively identified.

Insulation sheaths made of non-metallic polymers such as ethylene propylene rubber and silicone rubber can be conveniently installed on cable terminals through heat shrinkage or cold shrinkage and therefore are widely used in train cable terminals. The insulation structure of the ethylene propylene rubber cable terminal mainly consists of a composite shed, stress tube, and main insulation. An artificial test sample was made to simulate the real cable terminal for terahertz testing. All materials of this sample were taken from the actual cable terminal (QTO-J30G-25 kV) installed on the CHR380A Harmony high-speed train. The ST-I-CC structured sample includes three layers: copper conductor (inner layer), main insulation layer (middle layer), and stress tube (outer layer). To simulate the damp environment of the cable terminal, a rectangular area was set up on the main insulation layer, and pure water was injected. To reduce the attenuation of terahertz waves during refraction and reflection, the sample surface must be smooth and the thickness uniform. During the terahertz testing process, to reduce the interference of water vapor on the test results, high-concentration nitrogen was continuously injected into the test chamber, and the test environment temperature was maintained at around 25°C.

Firstly, damp insulation detection was conducted, and the terahertz frequency domain spectroscopy test was performed on the above-mentioned structural sample. As the damping conditions increased, the electric field strength decreased. The electric field strength at positions C-E in the undamped area was higher than at positions A and B. The frequency range of the effective spectrum in the undamped area was between 0.2~1.18 THz. In contrast, the effective frequency range of the damp area narrowed to between 0.2~0.75 THz. This is mainly due to the strong water absorption caused by the water filling in the damp area. To more clearly display the spectral changes caused by moisture, the linear spectrum was transformed into a logarithmic spectrum. The frequency domain spectral curves of the insulation damping are similar, but the curves at different measurement positions in the damping area are significantly different from the undamped area. The electric field strength at the undamped position is higher than in the damping area, with a frequency range of 0.2~2.8 THz. In the frequency range of 0.61~1.02 THz, the frequency domain spectrum at the undamped position is smoother. By comparing the frequency domain spectra of the undamped and damped areas, the damped and normal states can be distinguished.

By comparing the terahertz spectra of damp and undampened samples, we can preliminarily determine whether the composite insulation structure is damp. However, it is difficult to accurately determine the boundaries of the damping zone. This is mainly because each terahertz spectral curve only represents the damping situation at one measurement point of the sample. Therefore, terahertz spectral imaging tests are used to determine the damping area and its geometric shape. Different imaging methods are used, including frequency domain imaging, absorption spectral imaging, time domain maximum peak imaging, time domain minimum peak imaging, and peak-to-peak imaging. Terahertz frequency domain imaging can directly reveal the characteristic differences of different materials. The image directly reflects the intensity of the interaction between the material and the terahertz wave, and the image contrast is often used as an indicator to identify different materials. Since terahertz waves have different response frequencies for different types of materials, defects in the sample at the pixel can be reflected through terahertz frequency domain imaging. The moisture diffusion area has a clear boundary with the undamped area, making it easier to identify the actual wet area and the water diffusion area. However, due to the diffusion of moisture, the thickness of the affected area cannot be accurately displayed in the imaging results. Therefore, to compensate for the shortcomings of terahertz imaging, THz-TDS is used to determine the specific thickness of the damping area.

When the terahertz signal propagates from the pressure tube to the moisture area, it undergoes a precise process from optically dense to optically sparse materials, thereby generating a negative peak in the terahertz time-domain spectrum. Similarly, when the terahertz wave reaches the metal layer and water, a negative peak will appear in the terahertz time-domain spectrum, but the negative peak of the terahertz time-domain spectrum of the defect-free sample appears much later than that of the defective sample. After measurement, the average refractive index of the pressure tube is 2.91, and the average refractive index of the main insulation material is 1.61. The average thickness of the pressure tube calculated at positions A and B is 1.46 mm, and the average thickness of the pressure tube calculated at positions C to E is 1.60 mm. The test error of the thickness of the pressure tube at positions A and B is 2.7%, and the test error at positions C to E is 6.7%. Since water has a strong terahertz absorption, only the test results at positions C to E are used to calculate the thickness of the main insulation layer. The calculated average thickness is 1.53 mm, and the error with the actual value is only 2.0%. According to the test and calculation results, the depth of the damp layer of the composite cable terminal structure can be accurately determined.

At present, there is also technical limitation for non-destructive testing application of THz-TDS. For layered coatings, it is necessary to obtain the thickness of each layer through multiple measurements. An improved model method can be used, considering the roughness of the surface and interface for the measurement of the thickness of multilayer coatings [22]. Electromagnetic waves incident on a smooth surface is reflected in the mirror direction. When the surface is rough, the reflected wave will be diffusely scattered, thereby reducing the amplitude of the reflected wave. For any layered sample, it is necessary to follow the return path from the base to the incident medium and find the effective complex reflectance of each interface in turn. Using the effective complex reflectance of each layer, the terahertz radiation characteristics within any thin film are simulated, and the effective complex reflectance of the multilayer sample is derived. The reflected terahertz signal of the multilayer sample is simulated using the effective complex reflectance that depends on the parameters of each layer. By comparing the simulated signal with the measured signal of the sample, the unknown parameters of the coating can be determined in the numerical optimization step. At the same time, advanced signal processing methods for feature extraction with weak defect information are also an important direction for terahertz non-destructive testing. Using various algorithm models can further expand the application range of non-destructive testing of THz-TDS.

The cost of THz-TDS systems is relatively high, which also limits their widespread engineering applications. In addition, the low power of terahertz radiation is still a major problem limiting the application of non-destructive testing of THz-TDS technology, providing power for terahertz signals to increase the penetration depth of terahertz waves and improve the signal-to-noise ratio of the measurement signal. In addition, the spectral range of the THz-TDS system should be expanded, which can improve the spatial resolution of the image and further identify smaller sample defects. high power broadband terahertz radiation sources can penetrate thicker materials and further expand the application.

References

1. Ren G, Zhu Z, Zhang J, et al. Broadband terahertz spectroscopy of paper and banknotes. Opt Commun. 2020;475: 126267.
2. Liu L, Yu H, Zheng C, Ye D, He W, Wang S, Li J, Wu L, Zhang Y, Xie J, et al. Nondestructive thickness measurement of thermal barrier coatings for turbine blades by terahertz time domain spectroscopy. Photonics. 2023;10(2):105.
3. Fukuchi T, Fuse N, Okada M, et al. Measurement of refractive index and thickness of topcoat of thermal barrier coating by reflection measurement of terahertz waves. Electron Commun Jpn. 2013;96(12):37–45.
4. Fukuchi T, Fuse N, Okada M, et al. Topcoat thickness measurement of thermal barrier coating of gas turbine blade using terahertz wave. Electron Commun Jpn. 2014;189(1):1–8.
5. Li C, Cai J, Zhand Z, et al. Measurement of thermal barrier coating thickness of turbine blade based on THz-TDS. Infrared Millim Wave Terahertz Technol. 2021;8(11906):199–207. SPIE.
6. Burger R, Frisch J, Hubner M, et al. THz-TDS reflection measurement of coating thicknesses at non-perpendicular incidence: experiment and simulation. Sensors. 2021;21(10):3473.

7. Ye D, Wang W, Huang J, et al. Nondestructive interface morphology characterization of thermal barrier coatings using terahertz time-domain spectroscopy. Coatings. 2019;9:89.
8. Ye D, Wang W, Zhou H, et al. Characterization of thermal barrier coatings microstructural features using terahertz spectroscopy. Surf Coat Technol. 2020;394: 125836.
9. Tu W, Zhong S, Luo M, et al. Non-destructive evaluation of hidden defects beneath the multilayer organic protective coatings based on terahertz technology. Front Phys. 2021;9:304.
10. Su K, Shen Y, Zeitler JA. Terahertz sensor for non-contact thickness and quality measurement of automobile paints of varying complexity. IEEE Trans Terahertz Sci Technol. 2014;4(4):432–9.
11. Van MJLM, Frank A, Mass DJHC. Thickness sensor for drying paints using THz spectroscopy. Opt Express. 2021;29:7514–25.
12. Dong J, Locquet A, Citrin DS. Enhanced terahertz imaging of small forced delamination in woven glass fibre-reinforced composites with wavelet de-noising. J Infrared Millim Terahertz Waves. 2016;37(3):289–301.
13. Zhang D, Ren J, Li L, et al. Nondestructive testing of glass fiber honeycomb composites using terahertz technology. J Photonics. 2019;48(2):0212002.
14. Zhang JY, Ren JJ, Li LJ, et al. THz imaging technique for nondestructive analysis of debonding defects in ceramic matrix composites based on multiple echoes and feature fusion. Opt Express. 2020;28:19901–15.
15. Zefang W. Application of terahertz spectroscopy in qualitative and quantitative detection of drugs. Opt Instrum. 2021;43(6):1–5.
16. Alves-Lima D, Song J, Li X, et al. Review of terahertz pulsed imaging for pharmaceutical film coating analysis. Sensors. 2020;20:1441.
17. Wataru M, Hiroyuki Y, Yoshifumi K, et al. Applying terahertz technology for nondestructive detection of crack initiation in a film-coated layer on a swelling tablet. Results Pharma Sci. 2012;2:29–37.
18. Zhonghao Z, Hongwei M, Jianjun L, et al. Research on interface detection of composite insulators based on terahertz waves. China Electr Eng J. 2020;40(3):989–98.
19. Jiang H, Mei H, Bian X, et al. Detection of double-layer air gap defects based on terahertz imaging method. In: 2021 IEEE International Instrumentation and Measurement Technology Conference (I2MTC); 2021. pp. 1–5.
20. Li J, Yang L, He Y, et al. Terahertz nondestructive testing method of oil-paper insulation debonding and foreign matter defects. IEEE Trans Dielectr Electr Insul. 2021;28(6):1901–8.
21. Li S, Cao B, Kang Y, et al. Nonintrusive inspection of moisture damp in composited insulation structure based on terahertz technology. IEEE Trans Instrum Meas. 2021;70:1–10.
22. Cao B, Wang M, Li X, et al. Noncontact thickness measurement of multilayer coatings on metallic substrate using pulsed terahertz technology. IEEE Sens J. 2020;20(6):3162–71.

MIX
Papier aus verantwortungsvollen Quellen
Paper from responsible sources
FSC® C105338

If you have any concerns about our products,
you can contact us on
ProductSafety@springernature.com

In case Publisher is established outside the EU,
the EU authorized representative is:
Springer Nature Customer Service Center GmbH
Europaplatz 3, 69115 Heidelberg, Germany

Printed by Libri Plureos GmbH
in Hamburg, Germany